ANALYZING THE LARGE NUMBER OF VARIABLES IN BIOMEDICAL AND SATELLITE IMAGERY

ANALYZING THE LARGE NUMBER OF VARIABLES IN BIOMEDICAL AND SATELLITE IMAGERY

PHILLIP I. GOOD

WILEY

A JOHN WILEY & SONS, INC., PUBLICATION

Published by John Wiley & Sons, Inc., Hoboken, New Jersey
Published simultaneously in Canada

For general information on our other products and services or for technical support, please contact our Customer Care Department within the United States at (800) 762-2974, outside the United States at (317) 572-3993 or fax (317) 572-4002.

Wiley also publishes its books in a variety of electronic formats. Some content that appears in print may not be available in electronic formats. For more information about Wiley products, visit our web site at www.wiley.com.

Library of Congress Cataloging-in-Publication Data:

Good, Phillip I.
   Analyzing the large number of variables in biomedical and satellite imagery/Phillip I. Good.
      p. cm.
   Includes bibliographical references and index.
   ISBN 978-0-470-92714-4 (pbk.)
1. Data mining. 2. Mathematical statistics. 3. Biomedical engineering–Data processing. 4. Remote sensing–Data processing. I. Title.
   QA76.9.D343G753 2011
   066.3'12–dc22

                                                        2010030988

Printed in Singapore

   oBook ISBN: 978-0-470-93727-3
   ePDF ISBN: 978-0-470-93725-9
   ePub ISBN: 978-1-118-00214-8

10 9 8 7 6 5 4 3 2 1

CONTENTS

PREFACE

This text arose from a course I teach for http://statcourse.com on the specialized techniques required to analyze the very large data sets that arise in the study of medical images—EEGs, MEGs, MRI, fMRI, PET, ultrasound, and X-rays, as well as microarrays and satellite imagery.

The course participants included both biomedical research workers and statisticians, and it soon became obvious that while the one required a more detailed explanation of statistical methods, the other needed to know a great deal more about the biological context in which the data was collected.

Toward this end, the present text includes a chapter aimed at statisticians on the collection and preprocessing of biomedical data as well as a glossary of biological terminology. For biologists and physicians whose training in statistics may have been in a distant past, a glossary of statistical terminology with expanded definitions is provided.

You'll find that the chapters in this text are paired for the most part: An initial chapter that provides a detailed explanation of a statistical method is followed by one illustrating the application of the method to real-world data.

As a statistic without the software to make it happen is as useless as sheet music without an instrument to perform on, I have included links to the many specialized programs that may be downloaded from the Internet (in many cases without charge) as well as a number of program listings. As R is rapidly being adopted as the universal language for processing very large data sets, an R primer is also included in an appendix.

PHILLIP I. GOOD
HUNTINGTON BEACH CA
drgood@statcourse.com

CHAPTER *1*

VERY LARGE ARRAYS

1.1. APPLICATIONS

Very large arrays of data, that is, data sets for which the number of observations per subject may be an order of magnitude greater than the number of subjects that are observed, arise in genetics research (microarrays), neurophysiology (EEGs), and image analysis (ultrasound, MRI, fMRI, MEG, and PET maps, telemetry). Microarrays of as many as 22,000 genes may be collected from as few as 50 subjects. While EEG readings are collected from a relatively small number of leads, they are collected over a period of time, so that the number of observations per subject is equal to the number of leads times the number of points in time at which readings are taken. fMRI images of the brain can be literally four dimensional when the individual time series are taken into account.

Analyzing the Large Numbers of Variables in Biomedical and Satellite Imagery, First Edition.
Phillip I. Good.
© 2011 John Wiley & Sons, Inc. Published 2011 by John Wiley & Sons, Inc.

In this chapter, we consider the problems that arise when we attempt to analyze such data, potential solutions to these problems, and our plan of attack in the balance of this book.

1.2. PROBLEMS

1. The limited number of subjects means that the precision of any individual observation is equally limited. If n is the sample size, the precision of any individual observation is roughly proportional to the square root of n.

2. The large number of variables means that it is almost certain that changes in one or several of them will appear to be statistically significant purely by chance.

3. The large number of variables means that missing and/or erroneously recorded data is inevitable.

4. The various readings are not independent and identically distributed; rather, they are interdependent both in space and in time.

5. Measurements are seldom Gaussian (normally distributed), nor likely to adhere to any other well-tabulated distribution.

1.3. SOLUTIONS

Solutions to these problems require all of the following.

Distribution-free methods—permutation tests, bootstrap, and decision trees—are introduced in Chapters 2, 6, and 7, respectively. Their application to very large arrays is the subject of Chapters 3, 6, and 8.

One might ask, why not use parametric tests? To which Karniski et al. [1994] would respond:

Utilizing currently available parametric statistical tests, there are essentially four methods that are frequently used to attempt to answer the question. One may combine data from multiple variables to reduce the number of variables, such as in principal component analysis. One may use multiple tests of single variables and then adjust the critical value.

One may use univariate tests, and then adjust the results for violation of the assumption of sphericity (in repeated measures design). Or one may use multivariate tests, *so long as the number of subjects far exceeds the number of variables.*

Methods for reducing the number of variables under review are also considered in Chapters 3, 5, and 8.

Methods for controlling significance levels and/or false detection rates are discussed in Chapter 5.

Chapter 4, on gathering and preparing data, provides the biomedical background essential to those who will be analyzing very large data sets derived from medical images and microarrays.

PERMUTATION TESTS

Permutation tests provide *exact*, distribution-free solutions for a wide variety of testing problems. In this chapter, we consider their application in both two-sample single-variable and multivariable comparisons, in *k*-sample comparisons, in combining multiple single-variable tests, and in analyzing data in the form of contingency tables. Some R code is provided along with an extensive list of off-the-shelf software for use in performing these tests.

Their direct application to the analysis of microarrays and medical images is deferred to the next chapter.

2.1. TWO-SAMPLE COMPARISON

To compare the means of two populations, we normally compare the means of <u>samples</u> taken from those <u>populations</u>.[1] Suppose

---

[1]Definitions of underlined words will be found in the glossary of statistical terminology at the end of this book.

Analyzing the Large Numbers of Variables in Biomedical and Satellite Imagery, First Edition.
Phillip I. Good.
© 2011 John Wiley & Sons, Inc. Published 2011 by John Wiley & Sons, Inc.

our two samples consist of the observations 121, 118, 110, 34, 12, 22. Perhaps, I ought to indicate which observations belong to which sample, but if there really is no difference between the two populations from which the samples are drawn, then it doesn't matter how they are labeled. If I drew two equal sized samples, there are 20 possible ways the observations might be labeled as in the following table:

First Group	Second Group	Sum First Group
1. 121 118 110	34 22 12	349
2. 121 118 34	110 22 12	273
3. 121 110 34	118 22 12	265
4. 118 110 34	121 22 12	262
5. 121 118 22	110 34 12	261
6. 121 110 22	118 34 12	253
7. 121 118 12	110 34 22	251
8. 118 110 22	121 34 12	250
9. 121 110 12	118 34 22	243
10. 118 110 12	121 34 22	240
11. 121 34 22	118 110 12	177
12. 118 34 22	121 110 12	174
13. 121 34 12	118 110 22	167
15. 118 34 12	121 110 22	164
16. 110 34 12	121 118 22	156
17. 121 22 12	118 110 34	155
18. 118 22 12	121 110 34	152
19. 110 22 12	121 118 34	144
20. 34 22 12	121 118 110	68

If the null hypothesis were true, that is, if there really were no difference between the two populations, the probability that the observations 121, 118, and 110 might all be drawn from the first population by chance alone would be 1 in 20 or 5%. So to test if the means of two populations are the same:

1. Take two samples.
2. Consider all possible rearrangements of the labels of the two samples which preserve the sample sizes.
3. Compute the sum of the observations in the first sample for each rearrangement.
4. Reject the null hypothesis only if the sum we actually observed was among the 5% most extreme.

If you'd like to do a lot of unnecessary calculations, then instead of computing just the sum of the observations in the first sample, compute the difference in the two means, or, better still, compute *Student's t-statistic*. The denominator of the *t*-statistic is the same for each rearrangement as are the sample sizes as well as the total sum of all the observations, which is why the calculations are unnecessary.

Not incidentally, for samples of size 6 and above, you'd get approximately the same *p*-value if you computed the Student's *t*-statistic for the original observations and immediately looked up the result (or had your computer software look it up) in tables of Student's *t*. The difference between the two approaches is that the *significance level* you obtain from the *permutation test* described above is always *exact*, while that for Student's *t* is exact only if the observations are drawn from a *normal* (or Gaussian) distribution. Thankfully, the traditional Student's *t*-test is approximately exact in most cases as I have never encountered a normal distribution in real-life data.

2.1.1. Blocks

We can increase the sensitivity (*power*) of our tests by blocking the observations: for example, by putting all the men into one block and all the women into another,

$$\sum_{i=1}^{B}\sum_{j=1}^{n_i} x_{ij}$$

so that we will not confound the effect in which we are interested, for example, treatment, with an effect like gender, in which we are not interested. With blocked data, we rearrange the treatment labels separately within each block and then combine the test statistics with the formula, where B is the number of blocks and n_i is the sample size within the ith block.

2.2. k-SAMPLE COMPARISON

In the k-sample comparison, we have k sets of labels with n_i labels in the ith set. If we were to perform an analysis of variance, we would make use of the F-statistic.[2] The denominator of this statistic is the within sample variance, which remains unchanged when we rearrange the labels. Removing all other factors that remain unchanged, we are left with

$$F_2 = \sum_i \left(\sum_j X_{ij} \right)^2 / n_i$$

Let μ be the mean of the population from which the ith sample is drawn. Our null hypothesis is that $\mu_i = \mu$ for all samples. The alternative hypothesis is that one or more of the population means takes a different value from the others. If the losses we incur on deciding in favor of the null hypothesis when an alternative is true are proportional to the square of the difference, then the F-statistic or, equivalently, F_2 will minimize the losses.

---

[2]This statistic and other underlined words are defined in our glossary of statistical terminology.

But what if our loss function is not $\sum(\mu - \mu_i)^2$ but $\sum|\mu - \mu_i|$. Clearly, the optimal test statistic for discriminating between the null hypothesis and an alternative is

$$F_1 = \sum_{i=1}^{I} n_i|\overline{X}_{i.} - \overline{X}_{..}|$$

We cannot make use of this statistic if we use the analysis of variance to obtain a p-value; fortunately, its distribution, too, is easily determined by permutation means. Good and Lunneborg [2006] showed that the permutation test based on F_2 has the same power as the analysis of variance when the sample sizes are equal and is more powerful otherwise.

2.3. COMPUTING THE p-VALUE

Each time you analyze a set of data via the permutation method, you follow the same five-step procedure:

1. Analyze the problem—identify the alternative(s) of interest.

2. Choose a test statistic that best distinguishes between the alternative and the null hypothesis. In the case of the two-sample comparison, the sum of the observations in the sample from the treated population is the obvious choice as it will be large if the alternative is true and entirely random otherwise.

3. Compute the test statistic for the original labeling of the observations.

4. Rearrange the labels, then compute the test statistic again. Repeat until you obtain the distribution of the test statistic for all possible rearrangements or for a large random sample thereof.

5. Set aside one or two tails (i.e., sets of extreme values) of the resulting distribution as your rejection region. In the present example with a one-sided alternative, we would set aside the largest values as our rejection region. If the sum of the treatment observations for the original labeling of the observations is included in this region, then we will reject the null hypothesis.

For small samples, it would seem reasonable (and a valuable educational experience) to examine all possible rearrangements of labels. But for larger samples, even for samples with as few as six or seven values, complete enumeration may be impractical. While there are only 20 different ways we can apply two sets of three labels, there are 924 different ways we can apply two sets of six labels. Although 924 is not a challenge for today's desktop computers, if you are writing your own programs you'll find it simpler and almost as accurate to utilize the Monte Carlo method described next.

2.3.1. Monte Carlo Method

In a Monte Carlo method (named for the Monaco casino), we use the computer to generate a random rearrangement of the labels (control and treated; new and old). Suppose that **p** is the unknown probability that the value of the test statistic for the rearranged values will be as or more extreme than our original value. Then if we generate **n** random rearrangements, the number of values that are as or more extreme than our original value will be a binomial random variable $B(n, p)$ with mean np and variance $np(1 - p)$. Our estimate of p based on this random sample of n rearrangements will have an *expected* value of p and a standard deviation of $[p(1 - p)/n]^{1/2}$.

Suppose p is actually 4% and we examine 400 rearrangements; then 95% of the time we can expect our estimate of p to lie between

$0.04 - 1.96[0.04*0.96/400]^{1/2}$ and $0.04 + 1.96[0.04*0.96/400]^{1/2}$ or 2% to 6%. Increasing the number of rearrangements to 6400, we would get a p-value that is accurate to within half of 1%.

2.3.2. An R Program

```
rearrange = function (M, first, second){
#M is the number of rearrangements to be examined
# first is a vector holding the data from the first
  sample
# second holds the data from the second sample
n=length (second)
sumorig = sum(second)
cnt= 0 #zero the counter
#Stick both sets of observations in a single vector
data = c(first, second)
for (i in 1:M){
    D= sample (data,n)
    if (sum(D) <= sumorig) cnt=cnt+1
    }
cnt/M #pvalue
}
```

2.4. MULTIPLE-VARIABLE COMPARISONS

The value of an analysis based on simultaneous observations on several variables, such as height, weight, blood pressure, and cholesterol level, is that it can be used to detect subtle changes that might not be detectable, except with very large, prohibitively expensive samples, were we to consider only one variable at a time. The permutation test can be applied in a multivariate setting providing we can either

1. Find a single-valued test statistic that can stand in place of the multivalued vector of observations; or

2. Combine the *p*-values associated with the separate *univariate* tests into a single *p*-value.

2.4.1. Euclidean Distance Matrix Analysis

Consider an object represented by K landmarks in D dimensions. Such a biological object can be represented by a $K \times D$ matrix of landmark coordinates. The Euclidean distance matrix consisting of all interlandmark distances in two or more dimensions is coordinate system free and invariant under translation, rotation, and reflection. Fans of the TV show *Bones* will appreciate the depiction of skull landmarks in Figure 2.1.

For comparing two samples, Lele and Richtsmeier [1995] make use of a form ratio matrix whose elements are the ratio of the off-diagonal elements in the corresponding distance matrices. Its permutation distribution is readily calculated.

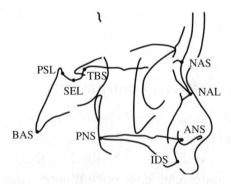

FIGURE 2.1 *Osseous*[3] landmarks located on the tracing of a head X-ray in two dimensions. All landmarks are located on the *sagittal* plane. Outline of the skin surface is shown for orientation. Reproduced with permission from John Wiley & Sons.

---

[3]Terms both italicized and underlined are defined in our glossary of biological terminology found at the end of this book.

2.4.2. Hotelling's T^2

The most commonly applied measure of the distance between two samples is Hotelling's T^2, a straightforward generalization of Student's t. Some notation is necessary, alas. We use bold type to denote a vector $\mathbf{X}$ whose components $(X_1, X_2, \ldots, X_J)$, have expected values $\boldsymbol{\mu} = (\mu_1, \mu_2, \ldots, \mu_J)$. Each component X_j corresponds to a different gene or a different point in an image. If we have collected vectors of observations for n subjects, then let $\overline{\mathbf{X}}$ denote the corresponding vector of sample means, and $\mathbf{V}$ the matrix whose ijth component is the covariance of X_i and X_j.

In the two-sample case, Hotelling's T^2 is defined as

$$(\overline{\mathbf{X}}_1 - \overline{\mathbf{X}}_2)V^{-1}(\overline{\mathbf{X}}_1 - \overline{\mathbf{X}}_2)^T$$

where the ijth component of $\mathbf{V}$ is estimated from the combined sample by the formula

$$V_{\ddot{g}} = \frac{1}{n_1 + n_2 - 2} \sum_{g-1}^{2} \sum_{k-1}^{k_R} (x_{gik} - \overline{x}_{gi})(x_{gik} - \overline{x}_{gi})$$

This statistic weighs the contribution of individual variables and pairs of variables in inverse proportion to their *covariances*.[4] This has the effect of rescaling each variable so that the most weight is given to (a) those variables that can be measured with the greatest precision and (b) those that can provide information not provided by the others. (See Figure 2.2.)

Unlike Student's t, the distribution of Hotelling's T^2 will depend on the nature of the distributions of the various variables. Thankfully, a permutation test based on Hotelling's T^2 is

---

[4]The covariance of two variables X and Y is related to the correlation ρ of X and Y by the formula $\text{Cov}(XY) = \text{Stdev}(X)\text{Stdev}(Y)\rho$.

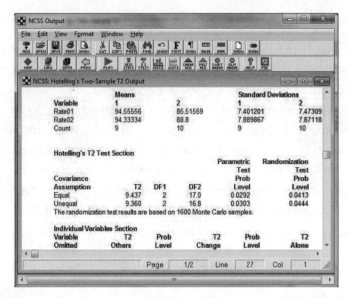

FIGURE 2.2   Results of a multivariate analysis using Hotelling's T^2. Note that p-values obtained from permutation distribution differ from those based on a bivariate normal.

distribution free. Using it, we can obtain *exact significance levels* and control the number of unexpected and unwanted errors. To perform this test:

1. Take two samples. Compute Hotelling's T^2.
2. Consider all possible relabelings of the two samples which preserve the sample sizes.
3. Compute Hotelling's T^2 for each rearrangement.
4. Reject the null hypothesis only if the sum we actually observed was among the 5% most extreme.

2.4.3. Mantel's *U*

A host of other <u>metrics</u> can be employed, such as the <u>Kolmogorov–Smirnoff</u> distance and Mantel's *U*.

Mantel's $U = \sum \sum a_{ij}b_{ij}$, where a_{ij} is a measure of the difference in time between items i and j, while b_{ij} is a measure of the spatial distance. As an example, suppose t_i denotes the time of the first passage over a predetermined threshold for the ith individual and (x_i,y_i) are the coordinates of the EEG lead. For all i, j set $a_{ij} = 1/(t_i - t_j)$ and

$$a_{ij}b_{ij} = 1/\sqrt{(x_i - x_j)^2 + (y_i - y_j)^2}$$

A large value for U would support the view that the impulse spreads methodically through the brain. How large is large? As always, we compare the value of U for the original data with the values obtained when we fix the i values but permute the j values as in $\pi[U] = \sum \sum a_{ij}b_{i\pi[j]}$.

2.4.4. Combining Univariate Tests

The statistical methods described in this section have the advantage that they apply to continuous, ordinal, or categorical variables or to any combination thereof. They can be applied to one-, two-, or k-sample comparisons.

Suppose we have made a series of vector-valued observations on K plants, each vector consisting of the values of J variables. The first variable might be a 0 or 1 according to whether or not the kth seedling in the ith treatment group germinated, the second might be the height of the kth seedling, the third the weight of its fruit; the fourth a subjective appraisal of fruit quality, and so forth. With each variable is associated a specific type of hypothesis test, the type depending on whether the observation is continuous, ordinal, or categorical and whether it is known to have come from a distribution of specific form. Let $\mathbf{T}_0$ denote the vector of single-variable test statistics derived from the original matrix of observations. These might include differences of means or weighted sums of the total number germinated, or any other statistic one might employ when testing just one variable at a time. When we rearrange the treatment labels on the observation vectors we obtain a new vector of single-variable test statistics $\mathbf{T}_\pi$.

In order to combine the various tests, we need to reduce them all to a common scale. Proceed as follows:

- Generate S rearrangements of X and thus obtain S vectors of univariate test statistics.
- Rank each of the single-variable test statistics separately. Let R_{ij} denote the rank of the test statistic for the jth variable for the ith rearrangement when it is compared to the value of the test statistic for the jth variable for the other S rearrangements including the values for the original test statistics.
- Combine the ranks of the various univariate tests for each rearrangement using Fisher's omnibus method:

$$U_i = -\sum_{j-1}^{J} \log \left[\frac{S + 0.5 - R_{ij}}{S + 1} \right]; \quad i = 1, \ldots, S$$

- Reject the null hypothesis only if the value of U for the original observations is an extreme value.

This straightforward, yet powerful method is due to Pesarin [2001]. Note that the tests may be dependent.

2.4.5. Gene Set Enrichment Analysis

We consider another multivariate method, specific to microarrays, in Section 5.4.

TABLE 2.1

	In Gene Set	Not in Gene Set	Marginal Total
Differentially Expressed	9	1	10
Not Differentially Expressed	4	10	14
Marginal Total	13	11	24

2.5. CATEGORICAL DATA

Sometimes the data falls in the form of a contingency table: for example, Table 2.1.

The 9 denotes the nine differentially expressed *genes*, which were in the gene set of interest; the 1 denotes the remaining differentially expressed genes, and so forth.

We see in this table an apparent difference in the expression rates of genes that were in and not in the data set of interest: 9 in 10 versus 4 in 14. How can we determine whether this difference is statistically significant?

The preceding contingency table has several fixed elements, its marginal totals:

- The total number of genes that were differentially expressed, 10.
- The total number of genes that were not, 14.
- The total number in the gene set, 13.
- The total number not in the gene set, 11.

These totals are immutable; no swapping of labels will alter the total number of individual genes or the numbers that were differentially expressed. But these totals do not determine the contents of the table, as can we seen in Tables 2.2 and 2.3, which have identical marginals.

Table 2.2 makes a strong case for the diagnostic value of the selected gene set, even stronger than our original observations.

TABLE 2.2

	In Gene Set	Not in Gene Set	Marginal Total
Differentially Expressed	10	0	10
Not Differentially Expressed	3	11	14
Marginal Total	13	11	24

TABLE 2.3

	In Gene Set	Not in Gene Set	Marginal Total
Differentially Expressed	8	2	10
Not Differentially Expressed	5	9	14
Marginal Total	13	11	24

In Table 2.3, whether or not a gene is in the designated set seems less important than in our original table.

These tables are not equally likely, not even under the null hypothesis. Table 2.2 could have arisen in any of 13 choose 3 ways, Table 2.3 of 13 choose 5 times 11 choose 2 ways.

Sir Ronald Fisher would argue that if the rate of differential expression were the same for all genes, then each of the redistributions of labels to subjects, that is, each of the N possible contingency tables with these same four fixed marginals, is equally likely, where[5]

$$N = \sum_{x=0}^{10} \binom{13}{x} \binom{11}{10-x} = \binom{24}{10}$$

How did we get this value for N? The component terms are taken from the *hypergeometric* distribution:

[5]Note that $\binom{24}{10}$ is interpreted as $24!/[10!(24-10)!]$.

$$\sum_{x=0}^{t} \binom{m}{x}\binom{n}{t-x} \bigg/ \binom{m+n}{t}$$

If the null hypothesis is true, then all tables with the marginals m, n, t are equally likely, and are as or more extreme. In our example, $m = 13, n = 11, x = 9$, and $t = 10$, so that of the N tables $\binom{14}{10}\binom{10}{1}$ are as extreme as our original table and $\binom{14}{11}\binom{10}{0}$ are more extreme. The resulting sum is still only a very small fraction of the total N, so we conclude that a difference in expression as extreme as the difference we observed in our original table is very unlikely to have occurred by chance. We reject the null hypothesis and accept the alternative hypothesis that the genes in the set we selected are more likely to be of diagnostic value.

2.6. SOFTWARE

Many years ago, when there were wolves in Wales, I attempted to calculate Hotelling's T^2 for some Hiroshima mortality data using a desktop calculator. This required that I pull a handle for each operation. I repeated the calculations for the original data three times, but never got the same answer. Today, I will not recommend a statistical procedure, no matter how extravagant the claims for it, unless computer software is available to perform the necessary calculations.

NCSS software performs the randomization test (permutation test) for Hotelling's T^2; see http://www.ncss.com/hotelt 2.html. Or download the software from http://www.stat.tamu .edu/~rfan/software.html/ that was used by Chen, Dougherty, and Bittner [1997] for "A genome-wide association scan for rheumatoid arthritis data by Hotelling's T^2 tests," which may be viewed at http://www.biomedcentral.com/1753- 6561/3/S7/S6#B11.

If you're familiar with R, permtest compares two groups of high-dimensional signal vectors derived from microarrays for a difference in location or variability. Download from `http://cran.r-project.org/web/packages/permtest/index.html`.

Blossom is a free interactive Windows program utilizing multiresponse permutation procedures for grouped data, agreement of model predictions, circular distributions, goodness of fit, least absolute deviation, and quantile regression. Download from `http://www.fort.usgs.gov/products/software/blossom/blossom.asp`.

To perform Pesarin's omnibus test, download a trial version of NPCtest from `http://www.gest.unipd.it/~salmaso/NPC_TEST.htm`.

A SAS macro to perform Pesarin's omnibus test may be downloaded from `http://homes.stat.unipd.it/pesarin/NPC.SAS`.

Almost every statistical package provides for Fisher's exact test. The program StatExact sold by cytel.com covers contingency tables with arbitrary numbers of rows and columns.

2.7. SUMMARY

In this chapter you were introduced to the application of distribution-free tests in hypothesis testing. You learned that you were free to choose the most powerful statistic for use in your application and were not limited in your choice by whether tables of the statistic's distribution were available.

In each application, the same five-step procedure is followed:

1. Determine the alternative(s) of interest.
2. Choose a test statistic.
3. Compute the value of this statistic for the observations.
4. Rearrange the sample labels among the observations.
5. Compute the value of the test statistic for the rearranged observations.

Steps 4 and 5 are repeated several hundred to several thousand times to determine the probability of observing a value of the test statistic by chance alone that was as extreme as the value that was observed.

As we shall see in the next chapter, we are free to choose the statistic that best discriminates between the null hypothesis and the alternative of interest, that is, the statistic that maximizes the power of the test while controlling the Type I error. To help us achieve this goal, the R code of Section 2.3.2 can be generalized as follows:

```
N=    #number of rearrangements to be examined
sumorig = statistic_of_your_choice(sample)
cnt= 0 #zero the counter
for (i in 1:N){
    D = sample (A,)
    if (statistic_of_your_choice (D)<= sumorig)
      cnt=cnt+1
}
cnt/N #pvalue
```

To learn more about permutation tests and their application, see Good [2005, 2006]. An R primer is provided in an appendix at the end of this book. Hotelling's T^2 was first described in Hotelling [1931]. Mantel's U was first described in Mantel [1967]. Pesarin's combination method and its applications are described in Pesarin [2001]. An alternative combination method is described by Potter and Griffiths [2006].

CHAPTER 3

APPLYING THE PERMUTATION TEST

In this chapter we examine a number of practical applications of the permutation method to the large arrays of data gathered in EEGs, MEGs, and microarrays.[6] In any such application, researchers have to make several decisions.

1. How and whether to reduce the vast of amount of data they have gathered. In particular, should all variables be considered or only a subset?

2. Whether to make a series of single-variable comparisons or one multivariate comparison.

3. If a series of single-variable comparisons, then how to control the overall error rate?

4. Which test statistic to use.

5. If a multivariate comparison, should they use a single-valued statistic? Or an omnibus test? If a single-value

---

[6]The biological background of these methods is reviewed in Chapter 4.

Analyzing the Large Numbers of Variables in Biomedical and Satellite Imagery, First Edition.
Phillip I. Good.

statistic, should it be Hotelling's T^2, a related measure of distance, or some other summary statistic such as the arithmetic mean or the maximum value?

6. How to avoid confounding the effects of interest with other potential confounding variables such as gender and age.

7. The stage in the analysis at which the data should be permuted.

3.1. WHICH VARIABLES SHOULD BE INCLUDED?

Depending on the hypothesis under test, some genes are more diagnostic than others. For example, Czwan, Brors, and Kipling [2010] found significant correlations[7] between "fatty acid metabolism" with overall survival in breast cancer, as well as "receptor mediated endocytosis," "brain development," "apical plasma membrane," and "MAPK signaling pathway" with overall survival in lung cancer. Incorporating the effects of all genes, as in a distance function, introduces noise and reduces the diagnostic power of a test.

For microarrays, Goeman and Bühlmann [2007] propose the following variable-reduction procedure:

1. First, calculate a measure of differential expression for each gene. For example, a p-value from a t-test or some other statistical test for *differential expression*[8] of single genes.

2. Establish a cut-off value to separate differentially expressed from nondifferentially expressed genes. (See Chapter 5 for a discussion of possible methods.)

---

[7]Definitions of underlined words are found in the glossary of statistical terminology at the end of this book.

[8]Definitions of words that are both underlined and italicized are found in the glossary of biological terminology.

3. Construct a 2×2 contingency table with the four categories "differentially expressed (yes/no)" and "in the gene set (yes/no)."

4. Analyze the table using Fisher's exact test. Eliminate nondifferentially expressed genes from further consideration *only if* the result is statistically significant.

This approach has at least three shortcomings as noted in Tian et al. [2005] and Draghici et al. [2003]. First, only the most significant portion of the gene list is used to compute the statistic, treating the less-relevant genes as irrelevant. Second, the order of genes on the significant gene list is not taken into consideration. Simply counting the number of gene set members contained in the short list leads to loss of information, especially if the list is long and the difference between the more significant and the less significant is substantial. Third, the correlation structure of gene sets is not considered.

The null hypothesis may be formulated in either of two ways:

1. The genes in a selected gene set show the same pattern of associations with the phenotype of interest compared with the rest of the genes.

2. The selected gene set does not contain any genes whose expression levels are associated with the *phenotype* of interest.

To test these hypotheses, Tian et al. [2005] propose that we first put the data in a matrix whose rows are genes and whose columns are samples. Next, calculate a measure of association t_i between each gene i and the phenotype of interest. Form the statistic T by summing over all the measures of association t_i for the genes in the set of interest.

To test the first hypothesis, we form the permutation distribution by rearranging the rows of the data matrix and computing

T for each rearrangement. To test the second hypothesis, we rearrange the columns of the matrix each time.

The permutation distributions of different gene sets are not the same. To compare them requires a further normalization as described in Tian et al. [2005].

3.2. SINGLE-VALUE TEST STATISTICS

3.2.1. Categorical Data

Deviations from Hardy–Weinberg equilibrium can indicate inbreeding, population stratification, and associations among genes. To test for deviations, Wigginton, Cutler, and Abecasis [2005] suggest the use of the permutation distribution of the following statistic:

$$P_{\text{HWE}} = \sum_{n_{AB}^*} I[\Pr\{N_{AB} = n_{AB}|N, n_A\} \geq \Pr\{N_{AB} = n_{AB}^*|N, n_A\}]$$
$$= x\Pr\{N_{AB} = n_{AB}^*|N, n_A\}$$

$I[x]$ is an indicator function that is equal to 1 when the comparison is true and equal to 0 otherwise. The sum should be performed over all heterozygote counts n_{AB}^* that are compatible with the observed number of minor alleles, n_A. Code for calculating the permutation distribution in C/C++, R, and Fortran is available at http://www.sph.umich.edu/csg/abecasis/Exact/index.html.

3.2.2. A Multivariate Comparison Based on a Summary Statistic

Mootha et al. [2003] used microarray profiles of over 22,000 genes to explore gene expression for 43 males with different levels of glucose intolerance. As both a biologist and a statistician, I've learned it is essential that statisticians understand the biological rationale for an experiment before attempting an analysis. Accordingly, I quote at length from their article:

Microarray data can be used to classify individuals according to molecular characteristics and to generate hypotheses about disease mechanisms. This approach has been successful in the study of cancer, where large changes in the expression of individual genes have often been observed. When alterations in gene expression are more modest and a large number of genes tested, high variability between individuals and limited sample sizes typical of human studies make it difficult to distinguish true differences from noise.

Alterations in gene expression might manifest at the level of biological pathways or co-regulated gene sets, rather than individual genes. Subtle but coordinated changes in expression might be detected more readily by combining measurements across multiple members of each gene set. A straightforward strategy for identifying such differences is to examine top-ranking genes in a microarray experiment and then to create hypotheses about pathway membership. This is both subjective and post hoc and thus prone to bias. A more objective set of approaches tests for enrichment of pathway members among the top-ranking genes in a microarray study, comparing them to a null distribution in which genes are randomly distributed. Because functionally related genes are often co-regulated, a positive result in such a test can be due solely to intrinsic correlation in gene expression rather than any relationship between expression of pathway members and the phenotype of interest.

We rank all genes according to the difference in expression (using an appropriate metric, such as signal-to-noise ratio, SNR). The primary hypothesis is that the rank ordering of the genes in a given comparison is random with regard to the diagnostic categorization of the samples. The alternate hypothesis is that the rank ordering of the pathway members is associated with the specific diagnostic criteria used to categorize the groups of affected individuals.

We then measure the extent of association by a non-parametric, running sum statistic termed the enrichment score (ES) and record the maximum ES (MES) over all gene sets in the actual data from affected individuals. To assess the statistical significance of the MES, we use permutation testing of the diagnostic labels of the individuals. Specifically, we compare the MES achieved in the actual data to that seen in each of 1,000 permutations that shuffled the diagnostic labels among the samples.

In short, these authors reduced the amount of data by computing an enrichment score, then computed the permutation distribution of a single statistic, the maximum enrichment score.

3.2.3. A Multivariate Comparison Based on Variants of Hotelling's T^2

Hayden, Lazar, and Schoenfeld [2009] assessed statistical significance in microarray experiments using the distance between the arrays as measured by variants of Hotelling's T^2. Their techniques can be applied to the entire _genome_ or just to subsets of interest.

The gene expression values are represented by two groups of column vectors: $X_{11}, \ldots, X_{1N}$ and $X_{21}, \ldots, X_{2M}$, for groups 1 and 2, respectively, where N and M are the number of arrays in groups 1 and 2, respectively. Let $D[X_{ij}, X_{km}]$ be the dissimilarity between two signals. The test statistic D^* whose permutation distribution is to be determined is equal to $D_{12} - (D_{11} + D_{22})/2$, where

$$D_{12} = \frac{1}{NM} \sum_{i \leq N, j \leq M} D[X_{1i}, X_{2j}]$$

$$D_{11} = \frac{1}{N(N-1)} \sum_{i < j \leq N} D[X_{1i}, X_{1j}]$$

$$D_{22} = \frac{1}{M(M-1)} \sum_{i<j\leq M} D[X_{2i}, X_{2,j}]$$

The authors also provide formulas for use with paired data (before and after comparisons) and blocked data. As in Section 2.2.1, the latter test statistic is obtained by summing the test statistics for the various blocks.

R code for the principal test they employ may be downloaded from http://cran.r-project.org/web/packages/permtest/index.html. Or one can make use of the R package globaltest with the instructions

```
> source("http://bioconductor.org/biocLite.R")

>biocLite("globaltest")
```

See Figure 3.1.

3.2.4. Adjusting for Covariates

Shapleske et al. [2002] studied 72 men with schizophrenia. Of these, 41 had a prominent history of auditory–verbal hallucinations and 31 had no such history. The patients were compared with 32 age, gender, handedness, and IQ matched healthy controls.

An analysis of covariance (ANCOVA) model was then fitted at each *voxel* in standard space, where there were N proportional volume (probability) estimates for each tissue class. The model is written below with *grey matter* proportional volume as the dependent variable:

$$G_{kj} = m + a_k + \text{Age}_{kj} + \text{Hand}_{kj} + e_{kj}$$

Here, G_{kj} denotes the proportional volume of grey matter estimated at a given voxel for the jth individual in the kth group; m

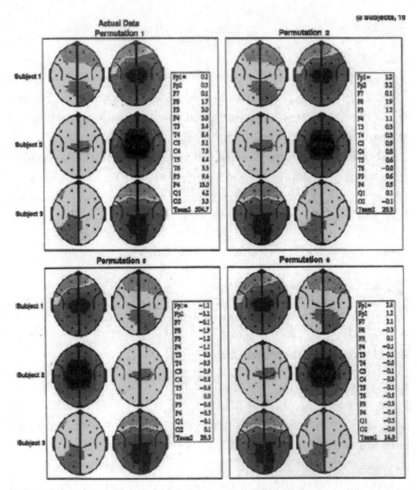

FIGURE 3.1 Possible relabelings of the outcomes. Reproduced from Karniski, Blair, and Snider [1994] with permission of Springer Science+Business Media.

is the overall mean; $m + a_k$ is the mean of the kth group; and e_{kj} is random variation. The independent variables Age_{kj} and $Hand_{kj}$ denote the age and handedness of the jth individual in the kth group.

Treating each Age × Hand combination as a block, it would then be appropriate to make several thousand random permutations among groups and within blocks to see at which voxels there were statistically significant differences.

3.2.5. Pre–Post Comparisons

MEG[9] data are collected as a set of N stimulus-locked event-related epochs (one per stimulus repetition) each consisting of a pre- and post-stimulus interval of equal length. Each epoch consists of an array of data representing the measured magnetic field at each sensor as a function of time. A *cortical* map is computed by averaging over all N epochs and applying a linear inverse method to produce an estimate of the temporal activity at each surface element in *cortex*.

Suppose the goal is to detect the locations and times at which activity during the post-stimulus experiment period differs significantly from the background pre-stimulus period. To apply a permutation test, we must find permutations of the data that satisfy an exchangeability condition, that is, permutations that leave the distribution of the statistic of interest unaltered under the null hypothesis. Permutations in space and time are not useful for these applications because of spatiotemporal dependence of the noise.

Instead, Pantazis et al. [2003] rely on the exchangeability of the pre- and post-stimulus data for each epoch. Given N original epochs, they create $M \leq 2N$ permutation samples, each consisting of N new epochs. Since the inverse operator is linear, one can apply the inverse before or after averaging the permuted epochs.

The authors began by summarizing the information in a series of steps as shown in Figure 3.2. Note that at each stage, the choice

---

[9]See Chapter 4 for a description of MEG data.

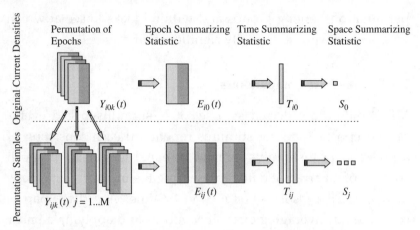

FIGURE 3.2 Illustration of the summarizing procedure used to construct empirical distributions from the permuted data: M permutation samples $Y_{ijk}(t)$ are produced from the original data $Y_{i0k}(t)$. The data are then summarized successively in epochs, time, and space to produce S_j. The empirical distribution of S_j can be used to draw statistical inferences for the original data. Reproduced from Pantazis et al. [2003] with permission from Springer Science+Business Media.

arises between using an average or the maximum with which to summarize the data. Statistical significance is determined by comparing S_0 with its permutation distribution, the $\{S_j\}$.

Pantazis et al. [2005] found that permutation methods had the same overall error rate as random field methods with smoothed data but were superior otherwise.

3.2.6. Choosing a Statistic: Time-Course Microarrays

Suppose that subject $i(i = 1; \ldots; n)$ contributes gene expression levels on J genes $(y_{i1}; \ldots; y_{iJ})$ at times $t_k(k = 1, \ldots, K)$. For gene $j(j = 1; \ldots; J)$, consider a time trajectory model such that the expected value $E(y_{ij}|t) = \mu_j(t)$.

Our null hypothesis for each gene j is that $\mu_j(t) = \mu$ for all t, while our alternative hypothesis is that the expression of this gene is a changing function of time, $\mu_j(t) \neq \mu_j(t')$, for some times t and t'.

We may treat this problem as if it were a one-way analysis with K groups (corresponding to the K points in time) with I subjects in each group. As a test statistic, we may use any of the statistics described in Section 2.2, for example,

$$\sum_{k=1}^{K} \left| \sum_{i=1}^{I} X_{ij}(t_k) \right|$$

$$\sum_{k=1}^{K} \left(\sum_{i=1}^{I} X_{ij}(t_k) \right)^2$$

or, if you anticipate the change will be strictly increasing or decreasing,

$$\sum_{k=1}^{K} t_k \sum_{i} X_{ij}(t_k)$$

Sohn et al. [2009] and Storey et al. [2005] use a quite different statistic based on spline estimates of the expected values $\{\mu_j(t)\}$. To determine whether a gene has expression that changes with age, they test whether a gene's population average time curve is flat. Figure 3.3 shows the expression measurements from a highly significant gene, CRABP1, a cellular retinoic acid-binding protein. They first fit a model under the null hypothesis of no differential expression, and then under the alternative hypothesis that there is differential expression. The null model is the dashed flat line in the figure that minimizes the sum of squares among all possible flat lines. The alternative model is the solid curve that minimizes the sum of squares among natural cubic splines.

Their objective is to discover genes whose expression levels are time dependent. That is, for each gene j, they test the hypothesis that the expected value of its expression is constant over time

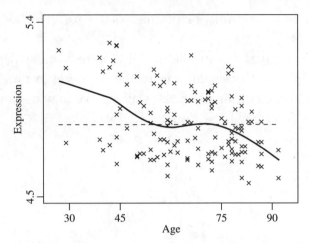

FIGURE 3.3 Expression values of a significant gene in the kidney aging study. The solid line is the curve fit under the alternative hypothesis of differential expression, and the dashed line is the curve fit under the null hypothesis of no differential expression. The × symbols represent observed expression values. Reproduced from Storey et al. [2005]. Copyright © 2005 National Academy of Sciences, U.S.A.

against the alternative hypothesis that the expected value varies at some point in time. Their test statistic is

$$F_j = (SSE_j^A - SSE_j^H)/SSE_j^A$$

where the superscripts A and H designate the sums of squares for error under the alternative hypothesis and the null hypothesis, respectively.

Sohn et al. [2009] used permutations of the measurement times to establish statistical significance; as they had to perform a large number of tests (one for each gene) they made use of a procedure we shall describe in Chapter 5 for controlling the overall error rate. Storey et al. [2005] used the bootstrap to test for significance and we shall examine their work a second time in Chapter 6.

3.3. RECOMMENDED APPROACHES

For EEGs, Wheldon, Anderson, and Johnson [2007] recommend the following four-step procedure:

1. Compute univariate statistics for all leads. Note that this statistic need not have a well-tabulated distribution, but may be any statistic that best discriminates among the various hypotheses.
2. Sum univariate statistics across leads.
3. Perform a permutation test for treatment effects at each time point.
4. Adjust for multiple comparisons (see Chapter 4).

For microarrays, we have the following procedure:

1. Compute univariate statistics for all genes. Note that this statistic need not have a well-tabulated distribution, but may be any statistic that best discriminates among the various hypotheses.
2. Rank all genes as to their diagnostic value, for example, via the magnitude of the associated univariate statistic. Select a diagnostic subset for further use.
3. Sum univariate statistics across genes in a subset.
4. Perform a permutation test.

3.4. TO LEARN MORE

Among the earliest applications of permutation tests to large arrays of data is described by Blair and Karniski [1992].

For a primer on the application of permutation methods to PET images with many worked-through examples, see Nichols and Holmes [2001, 2003]. See also Arndt et al. [1996]. Permutation methods are often applied to fMRI and MEG functional brain imaging (see Galán et al. 1997; Harmony et al. 2001; Suckling et al. 2006). Permutation methods also have been applied in structural brain imaging; see Thompson et al. [2001].

Discussion of hybrid methods, such as that of Belmonte and Yurgelun-Todd [2001], which delete less important voxels or genes as the algorithm progresses in order to control Type I error, are deferred to Chapter 5.

CHAPTER 4

BIOLOGICAL BACKGROUND

The objective of this chapter is to provide those whose primary training is in statistics with some of the biological background critical to understanding the limitations of the data they will be analyzing. Hopefully, our use of biomedical terminology in context will facilitate communication between the statistician and those he or she hopes to serve.

We examine in turn the data collection methods used in ultrasound, electroencephalograms (EEGs), and magnetoencephalograms (MEGs), nuclear magnetic resonance imaging (MRI), functional MRI, positron emission tomography, and microarrays.

4.1. MEDICAL IMAGING

Medical imaging entails various noninvasive procedures—MRI, ultrasound, and X-rays—for looking inside the bodies of humans and other animals.

Analyzing the Large Numbers of Variables in Biomedical and Satellite Imagery, First Edition. Phillip I. Good.

4.1.1. Ultrasound

A hand-held transducer acts as a speaker/transmitter 1% of the time as it projects high-frequency sound waves into the body and as a microphone/receiver 99% of the time as it listens for the echos. Solid bone reflects 100% of the sound (displayed as white on the viewing screen) while all of the sound passes through clear water (displayed as black on the viewing screen). Depending on their liquid content, the various soft tissues of the body appear on the viewing screen as shades of gray.

For computer-aided diagnosis, the analog signals from the VCR output of the scanner are transmitted to a frame grabber. The digital image is quantized into 8 bits (i.e., 256 gray levels) by using a software package with the frame grabber. The rectangular region of interest (ROI,) which extends beyond the lesion margins by 1–2 mm in all directions, is selected manually by a physician and saved as a file. See Figure 4.1.

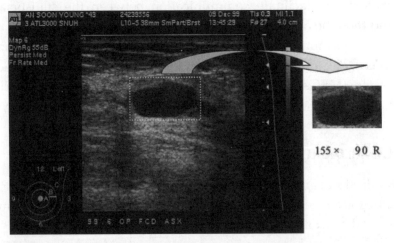

FIGURE 4.1 Digital image (640, 480 pixels) captured from the ultrasound (US) scanner. In a 1-cm × 1-cm rectangle, there are 94 × 94 = 8836 pixels. The ROI rectangle (arrow) is 1.65 cm × 0.96 cm and has 155 × 90 pixels. Reproduced from Kuo et al. [2002] with permission from Elsevier.

Possible sources of variation include the technologist who made the scan, the ultrasound unit, and the diagnosing physician.

Lumps are established as malignant or benign by means of fine-needle cytology, core-needle biopsy, or surgical biopsy, which is then reviewed by a pathologist/cytologist.

4.1.2. EEG/MEG

Electroencephalography and magnetoencephalography represent two noninvasive functional brain imaging methods, whose extracranial recordings, electroencephalograms (EEGs) and magnetoencephalograms (MEGs), measure electric potential differences and extremely weak magnetic fields generated by the electric activity of the neural cells, respectively. These recordings offer direct, real-time, monitoring of spontaneous and evoked brain activity and allow for spatiotemporal localization of underlying neuronal generators. Electroencephalography and magnetoencephalography are caused by the same neurophysiological events, that is, currents from synchronously activated neuronal tissue; thus both can be used for the localization of neuronal generators.

Electroencephalography and magnetoencephalography offer similar information about brain sources in what concerns accuracy of source localization, spatiotemporal resolution, and decoding or predictive power. The magnetoencephalography instrumentation costs about 20 times more than the electroencephalography instrumentation with the same number of channels. Still, both techniques are of value as the electroencephalography and magnetoencephalography measurement sensitivities are orthogonal. Electroencephalography primarily detects electric sources that are radial to the scalp surface with sufficiently distant electrodes and tangential components when the leads are located near to each other. Magnetoencephalography primarily senses magnetic currents generated by electric sources in the radial direction.

Preprocessing is essential (see Figure 4.2). The neuroelectric signals are buried in spontaneous EEGs with signal-to-noise ratios as low as 5 dB. In order to decrease the noise level and find a template evoked potential (EP) signal, an ensemble-average (EA) is obtained using a large number of repetitive measurements. This approach treats the background EEG as additive noise and the EP as an uncorrelated signal. The magnitudes and latencies of EP waveforms display large differences and changes depending on the psychophysiological factors for a given individual. Consequently, one goal in the methodological EP research is to develop techniques to extract the true EP waveform from a single sweep.

Electroencephalography has traditionally been measured using the standard 10–20 electrode system including only

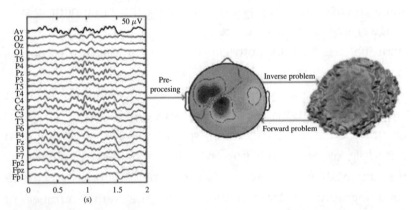

FIGURE 4.2 Key parts of source imaging. Preprocessing prepares the recorded signals for solving the inverse problem. The inverse problem attempts to locate the sources from recorded measurements, whereas the forward problem assumes a source definition in order to calculate a potential distribution map. Reproduced from Wendel et al. [2009] with permission.

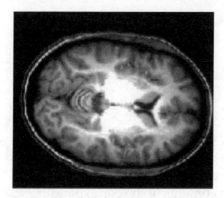

FIGURE 4.3 MRI of the brain distinguishes white and grey matter.

21 measurement electrodes. Commercially available systems including 128–256 electrodes are available today.

4.1.3. Magnetic Resonance Imaging

Nuclear magnetic resonance imaging (MRI) studies brain anatomy. Functional magnetic resonance imaging (fMRI) studies brain function.

4.1.3.1. MRI

As shown in Figure 4.3, the signal from hydrogen nuclei varies in strength depending on the surroundings. Collection of the reflected radiofrequency pulses provides us with a view within an otherwise impenetrable cranium. The nuclear magnetic relaxation times of tissues and tumors differ, so that MRI provides physicians with another noninvasive diagnostic tool (one far safer than X-rays).

As show in Figure 4.4, a set of radiofrequency pulses emitted by the MRI apparatus yield a free induction decay signal from the object being imaged in response. The intensity of a _pixel_ in the magnetic resonance image is proportional to the NMR signal

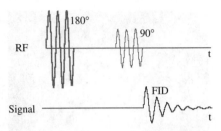

FIGURE 4.4   A set of radiofrequency pulses emitted by the MRI apparatus yield a free induction decay signal from the object being imaged.

intensity of the contents of the corresponding volume element or *voxel* of the object being imaged.

A variety of distinct pulse sequences are employed. Among the most often used are T_1- and T_2-weighted scans. T_1-weighted scans use a gradient echo sequence with a short echo time and a short repetition time. In the brain T_1-weighted scans provide good grey matter/white matter contrast. A T_2-weighted scan has a long echo time and a long repetition time. On a T_2-weighted scan, water- and fluid-containing tissues are bright and fat-containing tissues are dark; it is used to detect fluid accumulation indicative of injury or disease.

4.1.3.2. fMRI

Functional magnetic resonance imaging, or fMRI, is a technique for measuring brain activity. It works by detecting the changes in blood oxygenation and flow that occur in response to neural activity. When a brain area is more active, it consumes more oxygen; to meet this increased demand, blood flow to the active area increases. For example, when you move your right index finger there is a rapid momentary increase in the circulation of the specific part of the brain controlling that movement of the finger (see Figure 4.5).

The increase in circulation means an increase in oxygen, which is paramagnetic, and which affects the transmission times

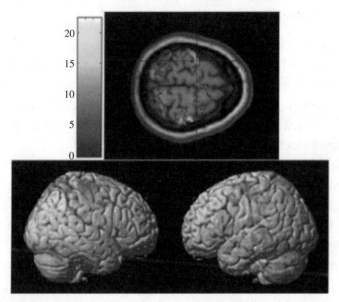

FIGURE 4.5 Example of functional imaging used to study the regions of the brain responsible for bilateral finger tapping. This image presents the regions of the brain experiencing the blood oxygen level dependent (BOLD) response. Used with permission from D.A. Kareken, Indiana University School of Medicine.

of the local brain tissues. The difference relative to surrounding tissues causes a contrast between the tissues, referred to as the blood oxygen level dependent (BOLD) response.

Imaging the brain during activity will elicit the BOLD response in those portions of the brain that are actively involved. The signal is very weak, requiring signal averaging. It is also a very rapid response and requires a fast imaging sequence. The time course of the signals need to be followed: Voxels whose signal corresponds closely to the activity are given a high activation score, voxels showing no correlation have a low score, and voxels showing the opposite (deactivation) are given a negative score. These scores can then be translated into the colors displayed on activation maps.

4.1.4. Positron Emission Tomography

Positron emission tomography (PET) imaging systems construct three-dimensional (3D) medical images by detecting gamma rays emitted when certain radioactively doped sugars are injected into a patient. Once ingested, these doped sugars are absorbed by tissues with higher levels of activity/metabolism (e.g., active tumors) than the rest of the body.

Gamma rays are generated when a positron emitted from the radioactive material collides with an electron in tissue. The resulting collision produces a pair of gamma-ray photons that emanate from the collision site in opposite directions and are detected by gamma-ray detectors arranged around the patient. Unlike anatomical imaging techniques like computed tomography (CT), X-ray, and ultrasound, PET imaging provides functional information about the human body.

The PET detector consists of an array of many thousands of scintillation crystals and hundreds of photomultiplier tubes (PMTs) arranged in a circular pattern around the patient. The scintillation crystals convert the gamma radiation into light, which is detected and amplified by the PMTs. To increase timing precision, the signals from a number (typically four or more) of physically close PMTs are summed, and this combined signal drives the input of an ultra-high-speed comparator (see Figure 4.6).

4.2. MICROARRAYS

A microarray consists of a set of probes, spotted on a solid surface, for example, a glass slide. Each probe is complementary to a specific fragment of one gene. A microarray can contain over six million probes, allowing for the expression of all genes to be analyzed simultaneously.

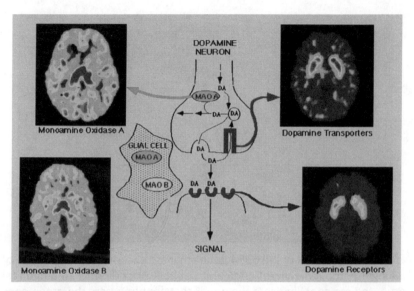

FIGURE 4.6 Diagram of the dopamine (DA) synapse along with the PET images for different molecular targets. Reproduced from Volkow, Rosen, and Farde [1997] through the courtesy of Brookhaven National Laboratory.

Each gene is responsible for the manufacture of a single protein. (A protein may be part of the structure of a cell or function as an enzyme or coenzyme that regulates the cell's metabolism.) The majority of the genes in the DNA molecule will be inactive for all or some portion of the cell's lifetime. When a gene is activated, it will serve as a template for messenger RNA (mRNA), which carries the information provided by the gene to the cell's cytoplasm, where the actual manufacture of proteins will take place.

To begin the analysis, all mRNA molecules (the primary products of gene expression) are extracted from the target samples. Then mRNAs are converted into complementary single-stranded DNA molecules (cDNA) in a process called reverse transcription. During this stage cDNAs are also labeled with fluorescent dyes

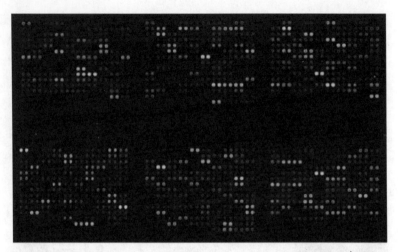

FIGURE 4.7   Most of the 6354 genes that are encoded in the genome of a fungal pathogen called *Candida albicans*. Responding to an antifungal drug, the red spots indicate the genes that are turned on, the green ones are turned off, and those that don't change are yellow. Reproduced with permission from Andre Nantel.

(e.g., cyanines). Sometimes, when the amount of biological material is very limited, cDNA amplification is required.

In the case of a two-color experiment, two samples—the control and the studied one—are prepared simultaneously. Each sample is labeled with a different dye (e.g., the control sample with Cy3, the examined sample with Cy5). Both samples are subsequently mixed and incubated overnight with the same microarray. During this stage, fluorescently tagged cDNAs hybridize with complementary probes.

Theoretically, each type of cDNA should only hybridize with one specific probe. Then the microarray is washed, dried, and scanned with two lasers. Each of them selectively detects signals coming from one of the applied fluorescent dyes. The intensity of the detected signal is directly proportional to the number of hybridized cDNA molecules, thus it also reflects the level of corresponding gene expression. (See Figure 4.7.)

As a result, two images are generated, the first showing the level of gene expression in the tested sample and the second in the control sample. The differences in the activity of the individual genes as well as the whole genomes can be superficially determined by superimposing both images. For a more precise, quantitative analysis, the images need to be converted to a numerical text data file. Finally, the raw data are submitted to preprocessing (normalization) procedures that minimize discrepancies between slides and allow for comparison of the results obtained from different experiments.

4.3. TO LEARN MORE

This chapter provided only a partial introduction to the methods used in collecting and preprocessing data from satellite and medical images and microarrays. For a background on the acquisition and preprocessing of satellite imagery, see `http://cast.uark.edu/home/research/environmental-studies/lulc-change-in-carroll-county/methodology1/data-acquisition.html`. To learn more about microarrays, see `http://pfgrc.jcvi.org/index.php/education/microarray_tutorial`.

CHAPTER 5

MULTIPLE TESTS

A realistic goal [of the statistician or bioinformatician] is to narrow the field for further analysis, to give geneticists a short-list of genes which are worth investing hard-won funds into analysing.

—MUKHERJEE [2003]

When we attempt multiple univariate (single-variable) tests of hypotheses rather than a single overall test, the resultant p-values may be dependent. Perhaps more crucial, if α is the significance level of each test ($\alpha = 0.05$ is typical), and we perform k tests, the probability P that at least one test will yield a p-value less than α by chance alone is given by the formula $1 - (1 - \alpha)^k$. If $\alpha = 0.05$ and k is 100, then P is greater than 99%!

In this chapter, we'll consider a variety of ways to control either the overall error or the false discovery rate.

Analyzing the Large Numbers of Variables in Biomedical and Satellite Imagery, First Edition.
Phillip I. Good.
© 2011 John Wiley & Sons, Inc. Published 2011 by John Wiley & Sons, Inc.

5.1. REDUCING THE NUMBER OF HYPOTHESES TO BE TESTED

The reduction proceeds in stages, first identifying which genes are differentially expressed and then selecting those of the greatest significance.

5.1.1. Normalization

Before the individual genes can be ranked according to their contributions or confidence intervals used to identify differentially expressed genes, the data must first be *normalized*.

Four normalization techniques are in common use. All four techniques assume that all (or most) of the genes in the array have an average expression ratio equal to one. The normalization factor is used to adjust the data to compensate for experimental variability and to balance the fluorescence signals from the samples being compared.

1. *Total intensity normalization* requires that the initial quantity of _mRNA_ be the same for both labeled samples, that for the thousands of genes in the array, approximately the same numbers in the query sample are upregulated as are downregulated relative to the control, and that the total integrated intensity computed for all the elements in the array should be the same in both the Cy3 and Cy5 channels.[10] A normalization factor can then be calculated and used to rescale the intensity for each gene in the array.

2. *Normalization using regression techniques.* In a scatterplot of the logarithms of Cy5 versus Cy3 intensities (or their logarithms), genes expressed at similar levels will cluster along a straight line. In closely related samples, the data

---

[10]See Chapter 4 for an explanation of these intensities. Words that are both underlined and italicized are defined in the glossary of biomedical terminology found at the end of this book.

can be normalized by adjusting the intensities so that the calculated slope is one.

(a) Although some authors, for example, Quackenbush [2001], recommend the use of local regression and other smoothing techniques, we do not, as the results while aesthetically pleasing have no causal basis; see Good and Hardin [2008].

3. *Normalization using ratio statistics.* Although individual genes may be up- or downregulated, in closely related cells, the total quantity of RNA produced is approximately the same for essential genes, such as "housekeeping genes." Chen, Dougherty, and Bittner [1997] used this fact to develop an approximate probability density for the ratio $T_k = R_k/G_k$, where R_k and G_k are, respectively, the measured red and green intensities for the kth array element, by normalizing the mean expression ratio to one.

4. Alternately, to equalize the variances of the intensities, one can make use of the logarithms of the intensities, or their cube roots. If the measurement error is proportional to the mean, then the log-transformed values will have consistent variance for all genes.

The Genomics and Bioinformatics group of NCI's Laboratory of Molecular Biology advises that for "studies where gene levels are fairly constant, and changes are expected to be modest, such as neuroscience studies, there is no need to transform data. For studies of cancerous tissue, where often some genes are elevated ten-fold or more, and these increases are highly variable between individuals, the log transform is very useful. For studies with at most moderate fold changes—such as most experimental treatments on healthy animals—it would be better to use a weaker transform, such as a cube-root transform or a 'variance-stabilizing' transform."

5.1.2. Selection Methods

Selection methods fall into two broad categories:

1. Computation of univariate statistics and rank.
2. Heuristic methods.

5.1.2.1. Univariate Statistics

The list of univariate statistics that have been employed to make pairwise comparisons is a lengthy one and includes the *t*-test, the Wilcoxon, empirical Bayes *t*-statistic, Welch *t*-statistic, fold change, and area under the receiver operating characteristic curve.

Fold Change. Fold change is the easiest method to implement, as at each locus one need only calculate the ratio of the intensities in the reference and experimental samples. These changes are then ranked and the highest ranking genes are selected.

t-Test. The *t*-test is also readily implemented while its more complex brethren—empirical Bayes *t*-statistic, signal-to-noise ratio, and the Welch *t*-statistic, used to correct for unequal variances—are difficult to justify.

Wilcoxon Rank. The Wilcoxon rank test makes less efficient use of the data but diminishes the effect of one or two large readings among the subjects. Whichever statistic is used, these are then ranked and the highest ranking genes are selected.

Kolmogorov–Smirnov Statistic. For gene j, nonparametric estimates of the gene expression distribution functions for diseased and healthy subjects are separately computed. The

Kolmogorov–Smirnov statistic is defined as the maximum distance between the two estimated distributions.

Cramér–von Mises Test. Let $x_1, x_2, \ldots, x_N$ and $y_1, y_2, \ldots, y_M$ be the observed values in the first and second sample, respectively, *in increasing order*. Let $r_1, r_2, \ldots, r_N$ be the ranks of the x's in the combined sample, and let $s_1, s_2, \ldots, s_M$ be the ranks of the y's in the combined sample. The test statistic is U defined as

$$U = N \sum_{i=1}^{N}(r_i - i)^2 + M \sum_{j=1}^{M}(s_j - j)^2$$

Bayesian Approach. In the *Bayesian* approach used by Long et al. [2001], the denominator employed in the t-statistic is estimated as if one had, in addition to the sample at hand, a second sample whose standard deviation is based on a local average of the standard deviations for genes showing similar within treatment expression levels to the gene under consideration. Local averaging is carried out by ordering all genes within a given treatment based on their average expression level and then taking the average standard deviation as the standard deviation observed for any given gene and the k next higher and lower expressing genes (where k is a user-defined constant).

Rank Product Method. The rank products method was developed by Breitling et al. [2004] for identifying differentially expressed genes.

Let $r[i, k]$ be the rank of the ith gene in the kth replicate, when the genes in each replicate are sorted by decreasing fold change. The rank product $R[i] = \prod r[i, k]$ if the number of genes in each replicate is the same, and $R[i] = \prod r[i, k]/n[k]$ otherwise.

Marginal Logistic Regression. We first normalize all genes to have unit variances. The marginal logistic regression for the *j*th gene assumes $\mathrm{logit}(E(Y = 1|X_j)) = \alpha_j + \beta_j X_j$, with unknown intercept α_j and regression coefficient β_j. T_j is set as the absolute value of the maximum likelihood estimate of β_j. Alternately, T_j is set as the *p*-value corresponding to the estimate of β_j from the logistic model.

Receiver Operating Characteristics (ROCs). Use of the ROCs was suggested by Parodi et al. [2003]. The area under the ROC curve (AUC) represents the probability that a randomly chosen positive example is correctly ranked with greater suspicion than a randomly chosen negative example. This probability is the same quantity estimated by the nonparametric Wilcoxon statistic.

Parodi et al. [2008] recommend the use of the area between the ROC curve and the rising diagonal instead as it can identify both proper (i.e., concave) and not proper ROC curves.

5.1.2.2. Which Statistic?

Jeffery, Higgins, and Culhane [2006] compared ten gene selection methods based on univariate statistics and found little agreement among the sets of genes each generated. They found that the area under a ROC curve performed well with data sets that had low levels of noise and large sample size. Rank products perform well when data sets had low numbers of samples or high levels of noise. Long et al. [2001] found that tests based on the analysis of variance and a Bayesian prior identify genes that are up- or downregulated following an experimental manipulation more reliably than approaches based only on a *t*-test or fold change. Using the bootstrap reproducibility index (see Chapter 6), Ma [2006] found considerable overlap among the genes selected by the various methods, although the overlaps are not perfect and the differences can be as large as 40%. The findings of Qiu et al. [2006] were similar to those of Ma. They report the most reliable results as those of the Cramér–von Mises test. The MAQC Consortium [2006] concluded that the fold change method showed the most reproducible results when intraplatform reproducibility for differently expressed genes was measured using the percentage of overlapping genes.

To automate the process of gene selection as well as to verify stability of the resulting lists, one may make use of the GeneSelector add-on package available without charge from the Bioconductor platform at http://www.bioconductor.org.

5.1.2.3. Heuristic Methods

As a mathematician, I have little use for heuristic methods except as a crutch one might make use of while trying to discern the underlying cause and effect relationships. Of course, all statistical methods have a heuristic aspect such as the arbitrary choice of the 5th percentile as a cutoff value in hypothesis testing. Still, the more complex the method, the greater my suspicions. In the words of John von Neumann, "with four parameters I can fit an elephant and with five I can make him wiggle his trunk."

SNPScan. According to Gresham et al. [2006], before the SNPScan algorithm can be employed some preliminary processing is required:

> Sequences were aligned and scanned for single *nucleotide* variants, identifying 86,443 potential single-nucleotide pairs (SNPs). Of these we eliminated any that were within 25 *base pairs* of another SNP, to ensure that we examined only *probes* that differed from the *hybridized* samples of DNA at a single base. To avoid confounding effects of inaccurate or incomplete sequence, we further eliminated any sites where the RM11–1a sequence had a *PHRED* score less than 30, as well as any within 50 base pairs of an *indel* in the RM11–1a/S288C alignment. In addition, we excluded probes that had more than 15 identical BLAST hits in the reference genome

and those probes for which the difference of the PM−MM ≤ -0.05 (9,563 of 2,464,467 probes across the whole array) in the analysis of the reference genome sample (strain FY3; see below). This left a training set of 24,848 SNPs, overlapped by 123,016 probes on the tiling array.

All intensity values were $\log_2$ transformed and normalized for the purpose of training the model and performing SNP predictions. To normalize each hybridization we used a method similar to Li and Wong's [2001] set-invariant normalization. We selected a subset of probes that appeared with high confidence to interrogate non-polymorphic regions of the sample. These were probes with intensity less than 1.5 standard deviations from their median value in five reference hybridizations. Use of this subset ensured that normalization eliminated only those intensity differences due to random experimental variation, and not true sequence differences between sample and reference. We then used a lowess [smoothing] procedure to fit the difference between sample and reference measurements as a function of sample intensity. We used this function to normalize the intensities of all probes on the sample microarray.

Gresham et al. modeled the difference between the experimental and the reference sample's intensities as

$$D_{ijt} = \alpha_{tj} + \beta_{tj}(GC_i) + \gamma_{tj}(PM_i - MM_i) + \delta_{tj}PM_i + I_{ijt}$$

where D_{ijt} is the signal difference for perfect match probe i overlapping a SNP at position j within the probe, with nucleotide triplet t including and flanking j, GC_i is the GC content of probe i, and PM_i and MM_i are the intensities of perfect match and mismatch probes for the reference sequence. I_{ijt} represents the two- and three-way interaction terms among the three covariates.

To check against <u>overfitting</u> due to the model's large size, they estimated its 4608 parameters[11] using only half of the probes

---

[11] Never mind fitting an elephant, with 4608 parameters they could fit a T-Rex.

in the training set and then used these values to predict signal changes for the other half. Observed and predicted values for the training set were correlated with an R^2 value of 0.73, while those for the test set were correlated with an R^2 of 0.68. They used the model to determine for each base in the genome the expected signal intensities of all overlapping probes under two distinct hypotheses: the base in the hybridized sequence is identical to the reference sequence or different from it. In the nonpolymorphic case, the expected intensity μ_{ni} of probe i was estimated as the observed mean intensity of i over the five reference genome's hybridizations. In the *polymorphic* case, the expected intensity μ_{pi} was estimated as $\mu_{ni} - D_{ijt}$. For both cases, the intensity variance was estimated based on the sample variance over the reference genome's hybridizations. They assigned each probe the average variance across a set of 501 probes with a similar minimum intensity.[12] Assuming a normal distribution in either case,[13] these values can then be used to generate a log likelihood ratio that the observed intensities x_i for the corresponding location k in a given sample indicates a polymorphic versus a nonpolymorphic site:

$$L_k = (\log_{10} e) \sum_i \frac{(x_i - \mu_{ni})^2 - (x_i - \mu_{pi})^2}{2\sigma_\iota^2}$$

(Note that they've taken the log of the likelihood ratio, summed over i probes overlapping the site, and for no obvious reason converted it to $\log_{10}$.) Positive values of L_k indicate a greater likelihood that site k is polymorphic, with higher values corresponding to increased confidence in this conclusion. This score is highly sensitive to underestimates of σ_i^2 if the observed intensity of a probe is far from either μ_{ni} or μ_{pi}. To avoid

---

[12]The authors did not explain why they chose 501.
[13]Something few would be willing to do.

misleading extremes introduced by this sensitivity, the authors increased the lowest variance among the probes overlapping a site to the value of the second-lowest variance[14], noting:

> In addition, probes were eliminated from the score calculation if they made a negative contribution to the prediction signal despite having intensities more than two standard deviations below μ_{ni}.

The authors did not test the consistency of the SNPSCan methodology.

ssG. ssGenotyping (ssG) is a multivariate, semisupervised approach proposed by Bourgon et al. [2009] for using microarrays to genotype haploid individuals. They found that ssG provides both more specific and more sensitive genotyping in the context of a meiotic recombination data set than does SNPScanner. Their results were not validated by use of the bootstrap.

Their method relies on the dubious premise that the data have a multivariate normal distribution. They fit this distribution via maximum likelihood estimates, a method for which I have nothing but contempt (see Good and Hardin [2008], p. 61).

Their algorithm, which maximizes an estimate of the conditional expectation of the log likelihood—only requires estimates of $\Pr\{Y_i = g|X_i\}$ for all $i \in S$. To initialize these conditional probabilities, they apply a clustering algorithm—k-means, with the two clusters seeded with parental observations—to the combined parental and segregant data, and then set each conditional probability to either 0 or 1, depending on the outcome of this clustering. The process being nonlinear, several iterations are required and are continued until a convergence criterion is met. The authors report a number of exceptions including (a) distributions that were not well separated by their method, (b) misclassifications, (c) outliers that needed to be removed, and (d) behavior that is

---

[14]Rather than the 3rd lowest or adding 19.5 to their wife's birthdate.

inconsistent with the biological and statistical models for meiotic recombination.

5.1.2.4. Which Method?

How might the authors of the latter two heuristic schemes have tested the consistency of their methodologies? Which approach is best for our application?

Several studies have examined feature selection by investigating the consistency between gene lists from small subsets of samples and those from the full data set, for example, Long et al. [2001], or using a bootstrap method to generate simulated datasets from real data sets, for example Pepe et al. [2003]. We address these questions in the next chapter when we consider methods of cross-validation.[15]

5.2. CONTROLLING THE OVERALL ERROR RATE

Troendle [1995] provides us with a resampling procedure that allows us to work around the dependencies. Suppose we have measured k variables on each subject and are now confronted with k test statistics. To make these statistics comparable, we need to standardize them and render them dimensionless, dividing each by its respective L_1 norm or by its standard error. For example, if one variable, measured in centimeters, takes values like 144, 150, and 156 and the other, measured in meters, takes values like 1.44, 1.50, and 1.56, we might divide each of the first set of observations by 6, and each of the second set by 0.06.

Next, we order the standardized statistics by magnitude, that is, from smallest to largest. We also reorder and renumber the corresponding hypotheses. The probability that at least one

---

[15]Underlined words are defined in the glossary of statistical terminology at the end of this book.

of these statistics will be significant by chance alone at the 5% level is $1 - (1 - 0.05)k$. But once we have rejected one hypothesis (assuming it was false), there will only be $k - 1$ true hypotheses to guard against rejecting.

To execute this step-down procedure:

1. Focus initially on the largest of the k test statistics, and repeatedly resample the data (with or without replacement) to determine the p-value.

2. If this p-value is less than the predetermined significance level, then accept this hypothesis as well as all the remaining hypotheses.

3. Otherwise, reject the corresponding hypothesis, remove it from further consideration, and repeat these steps.

5.2.1. An Example: Analyzing Data from Microarrays

The Monte Carlo permutation approach proposed by McIntyre et al. [2000] lends itself to the analysis of data from microarrays when we want to detect changes in up to 100,000 single nucleotide polymorphisms. For a collection of loci, there is a set of alleles[16] across all loci that is transmitted by the parents and a set of alleles across all loci that is not transmitted by the parents. Under the null hypothesis, the labels "transmitted" and "not transmitted" can be permuted for the sets of alleles. Let TDT denote the value of the test statistic for linkage at a single loci.

1. Calculate the maximum value of TDT—call this maximum TDTMAX—over all loci.

2. For each family, randomly exchange the "transmitted" and "not transmitted" labels.

---

[16] An allele is a variant of a gene, for example, the rh− and rh+ blood types result from two different alleles of the same gene.

3. Calculate TDTMAX for the data you've just permuted.

4. If TDTMAX for the permuted data is larger than the value of TDTMAX from the original data, count 1; otherwise count 0.

Repeat steps 2, 3, and 4 M times. Estimate the p-value as the total count from step 4 divided by the total number of rearrangements M.

5.3. CONTROLLING THE FALSE DISCOVERY RATE

Most of the procedures described previously are designed to control the familywise error rate (FWER). An alternate approach with large numbers of tests is to control the false discovery rate (FDR)—an approach first suggested by Benjamini and Hochberg [1995] and applied to microarrays by Drigalenko and Elston [1997], Reiner et al. [2003], and Korn et al. [2004]. When the tests are dependent, as is the case with microarrays, the subsampling approach proposed by Romano, Shaikh, and Wolf [2008] is often recommended. Their article is filled with technicalities, of great interest to mathematicians but of little practical import. I'd be happy to provide Dr. Wolf's C++ code for anyone who is interested.

Instead, let us consider, as an alternative, an adaptive step-down procedure due to Gavrilov, Benjamini, and Sarkar [2009] that is equally powerful, yet easier to follow and computationally far simpler.

Suppose we wish to control the false discovery rate at $q < 1$. First, sort the p-values corresponding to the m univariate hypotheses as $p(1) \leq \cdots \leq p(m)$.

Set the cutoff values at

$$c_i = iq/\{m + 1 - i(1 - q)\}, \quad i = 1, \ldots, m$$

Let $k = \max\{1 \leq i \leq m$ with $p(j) \leq c_j,\ j = 1, \ldots, i\}$. Reject the k hypotheses associated with $p(1) \ldots p(k)$ if k exists; otherwise accept all hypotheses.

```
#rcode. p is a vector of length m containing the
    p values.
CFDR=function(p, q){
m=length(p)
r=rank(p)
i=1
while (i<=m & p[m+1-r[i]] <= i*q/(m+1-i*(1-q)))i=i+1
return (i-1)
}
```

Note that we have not specified how the p-values were determined or what tests were employed as this will depend on your specific application.

5.3.1. An Example: Analyzing Time-Course Data from Microarrays

As we discussed in Section 3.2.5, Sohn et al. [2009] used permutations of the measurement times to establish statistical significance. To control the overall error rate, they used the following procedure:

1. Compute the m F-test statistics (one for each gene) from the original data, $F_{01}, \ldots, F_{0m}$.
2. For the bth permutation, compute the m F-test statistics $F_{b1}, \ldots, F_{bm}$.
3. Use a single-step procedure to control the overall error rate:
   (a) From the bth permutation data, calculate $u_b = \max_{1 \leq j \leq m} F_{bj}$.

(b) For gene j, calculate the adjusted p-value $p'_j = \sum_{b=1}^{B} I(u_b \geq F_{oj})/B$, where $I(\ )$ is an indicator function that is 1 if its argument is true and 0 otherwise.

(c) For a specified familywise error rate α, discover gene j if $p'_j < \alpha$.

A multiple testing procedure to control the FDR at α level can be obtained by replacing step 3 in the above algorithm with step 3' as described below:

3'. Use a multiple testing procedure:

(a) For gene j, estimate the marginal p-value by

$$p_j = \sum_{b=1}^{B} \frac{I(F_{bj} \geq F_{oj})}{B}$$

(b) For a chosen constant $0 \leq \lambda \leq 1$, estimate the q-value of gene j by

$$q_j = \frac{p_j \sum_{k=1}^{m} I(p_k > \lambda)}{(1 - \lambda) \sum_{k=1}^{m} I(p_k \leq p_j)}$$

(c) For a specified false discovery rate q^*, discover gene j if $q_j < q^*$.

See Figure 5.1.

5.4. GENE SET ENRICHMENT ANALYSIS

In this section, we discuss a multivariate procedure specific to the analysis of microarrays.

Typically, mRNA expression profiles are generated for thousands of genes from a collection of samples belonging to one of two _phenotypes_, for example, tumors that are sensitive versus

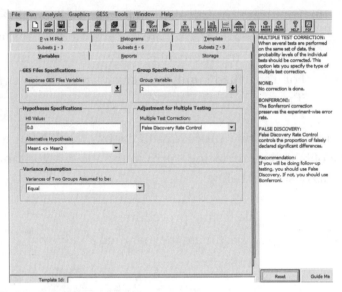

FIGURE 5.1 Setting GESS software to control the false discovery rate while performing multiple *t*-tests. See Section 5.5.

resistant to a drug. The genes can be ordered in a ranked list according to their differential expression between the classes.

Focusing solely on the handful of genes at the top and bottom of this list, that is, those showing the largest difference, has the following limitations.

1. After correcting for testing multiple hypotheses, no individual gene may meet the threshold for statistical significance.

2. Or, one may be left with a long list of statistically significant genes without any unifying biological theme.

3. Or, single-gene analysis may miss important effects on pathways. Cellular processes often affect sets of genes acting in concert. An increase of 20% in all genes encoding members of a metabolic pathway may dramatically alter the flux through the pathway and may be more important than a 20-fold increase in a single gene.

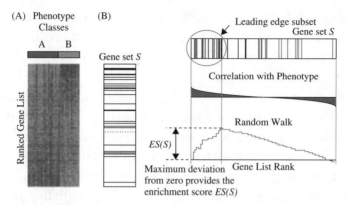

FIGURE 5.2 A GSEA overview illustrating the method. (A) An expression data set sorted by correlation with phenotype, the corresponding heat map, and the "gene tags," that is, location of genes from a set *S* within the sorted list. (B) Plot of the running sum for *S* in the data set, including the location of the maximum enrichment score (ES) and the leading-edge subset. Reproduced from Subramanian et al. [2005]. Copyright © 2005 National Academy of Sciences, U.S.A.

Given an a priori defined set of genes *S* (i.e., genes encoding products in a metabolic pathway, located in the same cytogenetic band, or sharing the same GO category), the goal of gene set enrichment analysis is to determine whether the members of *S* are randomly distributed throughout the ranked list or primarily found at its top or bottom.

Subramanian et al. [2005] calculate an enrichment score (ES) that reflects the degree to which a set *S* is overrepresented at the extremes (top or bottom) of the entire ranked list (Figure 5.2). The score is calculated by walking down the list, increasing a running-sum statistic when a gene in *S* is encountered, and decreasing it when we encounter genes not in *S*. The magnitude of the increment depends on the correlation of the gene with the phenotype. The enrichment score is the maximum deviation from zero encountered in the random walk; it corresponds to a weighted Kolmogorov–Smirnov-like statistic.

They determine the permutation distribution of ES by permuting the phenotype labels and recomputing the *ES* of the gene set for the permuted data. Permuting the class labels preserves gene–gene correlations, providing a more biologically reasonable assessment of significance than would be obtained by permuting genes.

When an entire database of gene sets is evaluated, the significance level derived from the permutation distribution must be revised to account for multiple hypothesis testing. We first normalize the ES for each gene set to account for the size of the set, yielding a normalized enrichment score (NES).We then control the proportion of false positives by calculating the false discovery rate (FDR) corresponding to each NES. The FDR is the estimated probability that a set with a given NES represents a false positive finding; it is computed by comparing the tails of the observed and null distributions for the NES.

1. Determine $ES(S)$ for each gene set in the collection or database.

2. For each S and 1000 fixed permutations π of the phenotype labels, reorder the genes in the ranked list and determine $ES(S, \pi)$.

3. Adjust for variation in gene set size. Normalize the $ES(S, \pi)$ and the observed $ES(S)$, separately rescaling the positive and negative scores by dividing by the mean of the $ES(S, \pi)$ to yield the normalized scores $NES(S, \pi)$ and $NES(S)$.

4. Compute the FDR. Control the ratio of false positives to the total number of gene sets attaining a fixed level of significance separately for positive (negative) $NES(S)$ and $NES(S, \pi)$.

Once this process is complete, create a histogram of all $NES(S,)$ over all S, and use this null distribution to compute an FDR q-value, for a given $NES(S) = NES^* \geq 0$. The FDR is the ratio of the percentage of all (S, π) with $NES(S, \pi) \geq 0$, whose

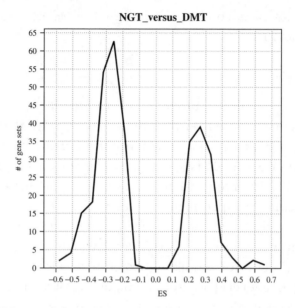

FIGURE 5.3 GSEA transcriptional profiles of smooth muscle biopsies of diabetic and normal individuals obtained by Mootha et al. [2003]; see Section 3.2.2.

$NES(S, \pi) \geq NES^*$, divided by the percentage of observed S with $NES(S) \geq 0$, whose $NES(S) \geq NES^*$, and similarly if $NES(S) = NES^* \geq 0$. (See Figure 5.3.)

To perform all the necessary calculations, *GSEA-P 2.0* is freely available for academic and commercial users and can be downloaded from `http://www.broadinstitute.org/gsea/msigdb /downloads.jsp`.

5.5. SOFTWARE FOR PERFORMING MULTIPLE SIMULTANEOUS TESTS

5.5.1. AFNI

AFNI incorporates permutation tests for use in analyzing neuroimages. Versions for Macs, PCs, and Unix computers may be downloaded from http://afni.nimh.nih.gov/afni/download/afni.

5.5.2. Cyber-T

Cyber-T is a statistics program with a web interface that can be conveniently used on high-dimensional array data for the identification of statistically significant differentially expressed genes. It employs statistical analyses based on simple t-tests that use the observed variance of replicate gene measurements across replicate experiments, or regularized t-tests that use a Bayesian estimate of the variance among gene measurements within an experiment. Cyber-T also contains a computational method (PPDE) for estimating experiment-wide false positive and negative levels based on the modeling of p-value distributions. Access it at http://cybert.microarray.ics.uci.edu/.

5.5.3. dChip

DNA-Chip Analyzer (dChip) is a Windows software package for probe-level (e.g., Affymetrix platform) and high-level analysis of gene expression microarrays and SNP microarrays using the methods of Li and Wong [2001] and Lin et al. [2004]. Gene expression or SNP data from various microarray platforms can also be analyzed by importing as external data set. At the probe level, dChip can display and normalize the CEL files, and the model-based approach allows pooling information across multiple arrays and automatic probe selection to handle cross-hybridization and image contamination. High-level analysis in dChip includes comparing samples, hierarchical clustering, view expression and SNP data along chromosome, LOH, and copy number analysis of SNP arrays, and linkage analysis. In these functions the gene information and sample information are correlated with the analysis results. Download from http://sites.google.com/site/dchipsoft/downloading-dchip-software.

5.5.4. ExactFDR

The ExactFDR software package improves speed and accuracy of the permutation-based false discovery rate (FDR) estimation method by replacing the permutation-based estimation of the null distribution by the generalization of the algorithm used for computing individual exact p-values. It provides a quick and accurate nonconservative estimator of the proportion of false positives in a given selection of markers. A Java 1.6 (1.5-compatible) version is available on SourceForge at http://sourceforge.net/projects/exactfdr.

5.5.5. GESS

GESS accepts a variety of file formats, provides for preprocessing and normalization, and has numerous hierarchical clustering algorithms, but lacks bootstrap to verify stability of findings. GESS controls the false discovery rate using the method of Benjamini and Hochberg [1995]. A seven-day trial may be downloaded from http://www.ncss.com/gess.html.

5.5.6. HaploView

This special purpose program designed for use with microarrays provides for permutation testing for association significance. It may be downloaded without charge from http://www.broadinstitute.org/mpg/haploview/index.php.

5.5.7. MatLab

MatLab code to perform gene selection using the method of Ai-Jun and Xin-Yuan [2010] is available without charge at http://www.sta.cuhk.edu.hk/xysong/geneselection/.

FieldTrip is a MatLab software toolbox for MEG and EEG analysis and includes cluster-based permutation tests. It is available at http://fieldtrip.fcdonders.nl/start.

5.5.8. R

permtest compares two groups of high-dimensional signal vectors derived from microarrays, for a difference in location or variability. You can download it from http://cran.r-project.org/web/packages/permtest/index.html.

multtest provides nonparametric bootstrap and permutation resampling-based multiple testing procedures (including empirical Bayes methods) for controlling the familywise error rate (FWER), generalized familywise error rate (gFWER), tail probability of the proportion of false positives (TPPFPs), and false discovery rate (FDR). Several choices of bootstrap-based null distribution are implemented (centered, centered and scaled, quantile-transformed). Single-step and stepwise methods are available. Tests based on a variety of t- and F-statistics (including t-statistics based on regression parameters from linear and survival models as well as those based on correlation parameters) are included. Results are reported in terms of adjusted p-values, confidence regions, and test statistic cutoffs. The procedures are directly applicable to identifying differentially expressed genes in DNA microarray experiments. To install this package for Windows or Macs, start R and enter:

```
source("http://bioconductor.org/biocLite.R")
biocLite("multtest")
```

5.5.9. SAM

SAM assigns a score to each gene on the basis of change in gene expression relative to the standard deviation of repeated measurements. For genes with scores greater than an adjustable threshold, SAM uses permutations of the repeated measurements to estimate the percentage of genes identified by chance, that is, the false discovery rate (FDR).

The input to SAM is gene expression measurements from a set of microarray experiments, as well as a response variable from each experiment. The response variable may be a grouping like untreated, treated (either unpaired or paired), a multiclass grouping (like breast cancer, lymphoma, or colon cancer), a quantitative variable (like blood pressure), or a possibly censored survival time. SAM computes a statistic d_i for each gene i, measuring the strength of the relationship between gene expression and the response variable. It uses repeated permutations of the data to determine if the expression of any gene is significantly related to the response. The cutoff for significance is determined by a tuning parameter delta, chosen by the user based on the false positive rate. One can also choose a fold-change parameter, to ensure that called genes change at least a prespecified amount.

SAM allegedly can be obtained from `http://www-stat-class.stanford.edu/~tibs/clickwrap/sam.html`, but we found that the page had errors.

SAM for use under R may be downloaded from *Linux/Unix version 1.25* or *Windows version 1.25*.

5.5.10. ParaSam

ParaSam is a parallel version of the SAM algorithm that executes remotely. Because the computing burden is placed on the remote computers rather than on the investigator's desktop, ParaSam is not only faster than the serial version of SAM, but can analyze extremely large data sets. ParaSam may be utilized from `http://bioanalysis.genomics.mcg.edu/parasam/`. See Table 5.1.

TABLE 5.1 Output from ParaSam

#	Rank	Gr1_Mean	Gr2_Mean	Gr2/Gr1	D-Statistic
1416076_at	1	5.9965	2.2680	0.3782	−29.60
1417153_at	2	2.5599	5.4661	2.1353	−24.68
1416258_at	3	7.0514	3.6093	0.5118	−23.35
1417911_at	4	9.3792	5.6282	0.6001	−21.48
1417910_at	5	8.8098	4.9432	0.5611	−19.86
1415874_at	6	7.0126	2.6292	0.3749	−17.61
1417458_s_at	7	10.9277	7.8390	0.7174	−16.74
1415810_at	8	10.3627	7.2157	0.6963	−16.01
1417823_at	9	3.7440	2.3738	0.6340	−15.95
1416120_at	10	7.6936	4.2029	0.5463	−15.85

5.6. SUMMARY

To correct for multiple hypotheses tests, one can either reduce the number of variables to be considered or attempt to control the overall Type I error or the false discovery rate. One cannot do both.

5.7. TO LEARN MORE

Celebi and Aslandogan [2004] provide methods for reducing the dimensionality via principal components analysis. Troyanska et al. [2002] compare various nonparametric methods for identifying differentially expressed genes in microarray data. Tusher, Tibshirani, and Chu [2001] describe the application of SAM to the significance analysis of microarrays.

Benjamini and Yekutieli [2001] describe methods for controlling the false discovery rate under dependency. The false discovery rate can be reduced by appropriate choice of the null hypothesis; see Efron [2004]. Methods for determining the approximate minimal sample size may be gleaned from Dobbin and Simon [2005] and Lin, Rogers, and Hsu [2001].

CHAPTER 6

THE BOOTSTRAP

Although the primary use of the bootstrap in the analysis of very large data sets is *validation*, it was first conceived of as a method for estimating precision, and later shown to be of value in such diverse applications as regression, sample-size determination, and comparing populations with unequal variances. In this chapter, we take a first look at all these applications.

In this chapter, you'll learn how to draw percentile, parametric, blocked, balanced, and adjusted bootstrap samples, and obtain R code that can assist you in the process. You'll see how the bootstrap was applied in gene screening, positron emission tomography, magnetoencephalography, and microarray analysis.

6.1. SAMPLES AND POPULATIONS

The best way to establish the values of *population parameters*—the population mean, its dispersion, and its various percentiles—is to examine each and every member of the population, make

Analyzing the Large Numbers of Variables in Biomedical and Satellite Imagery, First Edition.
Phillip I. Good.
© 2011 John Wiley & Sons, Inc. Published 2011 by John Wiley & Sons, Inc.

the appropriate measurement or measurements, and record their values. Alas, we seldom have the opportunity to conduct a complete census.

The method of examination may be destructive, as is the case when a can needs to be opened to check for contamination or a condom burst to measure its tensile strength.[17]

Or, the method of examination is prohibitively expensive or time consuming or both.

Or, the population is hypothetical in part; for example, when we test a new infant formula, we want to extrapolate our findings to all the children yet to be born.

In such cases, we select a sample of the members of the population and use the sample statistics to estimate the population parameters. For the population mean, use the sample mean; for the population median, use the sample median; and to estimate a proportion in the population, use the proportion in the sample.

But the sample is not the population. If we had taken a different sample, we might well have recorded quite different values. The best way to find out how different and thus how precise our estimates are is by taking repeated samples from the same population. Still, if we could have afforded this many samples, we would have taken a single very large sample in the first place, as estimates from larger samples are always more precise. (Note: Unbiased estimates from larger samples are also more accurate.)

6.2. PRECISION OF AN ESTIMATE

One practical alternative, known as the bootstrap, is to treat the original sample of values as a stand-in for the population and to

---

[17] At least two court cases testify to the reluctance of manufacturers to do just that: *Anderson & Co v. U.S.*, 284 F. 542,543 (9th Cir. 1922) and *U.S. v. 43 1/2 Gross Rubber Prophylactics*, 65 F. Supp. 534 (Minn. 4th Div. 1946).

Accuracy and Precision

Let us suppose Robin Hood and the Sheriff of Nottingham engage in an archery contest. Each is to launch three arrows at a target 50 meters (half a soccer pitch) away. The Sheriff launches first and his three arrows land one atop the other in a dazzling display of shooting *precision*. Unfortunately, all three arrows penetrate and fatally wound a cow grazing peacefully in the grass nearby. The Sheriff's *accuracy* leaves much to be desired.

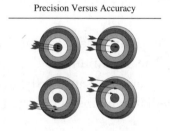

Precision Versus Accuracy

---

resample from it repeatedly, with replacement, computing the desired estimate each time.

Consider the following set of 22 observations on the heights of sixth-grade students, measured in centimeters and ordered from shortest to tallest. Note that the median of the sample, a plug-in estimate of the median of the population from which it is drawn, is 153.5 cm.

137.0 138.5 140.0 141.0 142.0 143.5 145.0 147.0 148.5 150.0 153.0

154.0 155.0 156.5 157.0 158.0 158.5 159.0 160.5 161.0 162.0 167.5

Suppose we record each student's height on an index card, 22 index cards in all. We put the cards in a big hat, shake them up, pull one out and make a note of the height recorded on it. We *return the card to the hat* and repeat the procedure for a total of 22 times until we have a second sample, the same size as the original. Note that we may draw a specific student's card several times as a result of using this method of sampling with replacement.

As an example, our first bootstrap sample, which I've arranged in increasing order of magnitude for ease in reading, might look like this:

138.5 138.5 140.0 141.0 141.0 143.5 145.0 147.0 148.5 150.0 153.0

154.0 155.0 156.5 157.0 158.5 159.0 159.0 159.0 160.5 161.0 162.0

Several of the values have been repeated as we are sampling with replacement. The minimum of this sample is 138.5 cm, higher than that of the original sample, the maximum at 162.0 cm is less, while the median remains unchanged at 153.5 cm.

137.0 138.5 138.5 141.0 141.0 142.0 143.5 145.0 145.0 147.0 148.5

148.5 150.0 150.0 153.0 155.0 158.0 158.5 160.5 160.5 161.0 167.5

In this second bootstrap sample, we again find repeated values; this time the minimum, maximum, and median are 137.0 cm, 167.5 cm, and 148.5 cm, respectively.

The bootstrap can be used to determine the precision of any estimator. For example, the variance of our sample of heights is 76.7 cm^2. The variances of 100 bootstrap samples drawn from our sample range between 47.4 cm^2 and 115.6 cm^2 with a mean of 71.4 cm^2. They provide a feel for what might have been had we sampled repeatedly from the original population. The resulting values from our 100 bootstrap samples are summarized in Figure 6.1.

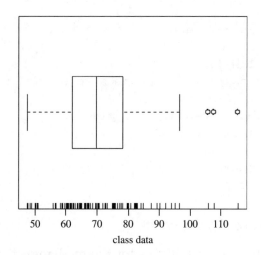

class data

FIGURE 6.1 Boxplot and strip chart of variances of 100 bootstrap samples.

6.2.1. R Code

```
#This program selects 50 bootstrap samples from the
 classroom data
#and then produces a boxplot and stripchart of the
 results.
class=c(141,156.5,162,159,157,143.5,154,158,140,
 142,150,148.5,138.5,161,153,145,147,158.5,160.5,
 167.5,155,137)
#Record group sizes
n = length(class)
#set number of bootstrap samples
N =50
stat = numeric(N) #create a vector in which to
 store the results
#the elements of the vector will be numbered from 1
 to N
```

```
#Set up a loop to generate a series of bootstrap
 samples
for (i in 1:N){
#bootstrap sample counterparts to observed
 samples are denoted with ''B''
classB= sample (class, n, replace=T)
stat[i] = median(classB)
}
boxplot (stat)
stripchart(stat)
```

6.2.2. Applying the Bootstrap

By substituting other statistics for the median in the preceding program, one may obtain bootstrap estimates of precision for all of the following statistics.

1. *Percentiles* of the population's *frequency distribution*.
2. Central values such as the *arithmetic mean*, the *geometric mean*, and the *median*.
3. Measures of dispersion such as the *variance*, the *standard deviation*, and the *interquartile deviation*.
4. Or, as in an audit I performed, the total expenditure $\sum f_i c_i$, where c_i is the cost of the ith item and f_i the frequency with which it occurs.

With the aid of the bootstrap and the permutation test, we are able to choose the best statistic for an application; no longer are we limited by the availability of tables. Here is one example. Suppose we have a series of independent identically distributed (i.i.d.) observations with cumulative distribution $\Pr\{X_i \leq x\} = F[x - \Delta]$ and we want to estimate the location parameter Δ without having to specify the form of the distribution F. If F is the <u>normal</u> or Gaussian distribution, the location parameter Δ corresponds to both the population mean and the population median. And if the loss function is proportional to the square of the estimation error, then the arithmetic mean of the sample is optimal for estimating Δ.

Suppose, on the other hand, that F is symmetric just as the normal distribution is, so that the population mean is still equal to the population median, but F may include many more very large or very small values than a normal distribution would. Then the sample median is to be preferred as an estimate of Δ.

If we are uncertain whether or not F is symmetric, then our best choice of an estimator is a statistic that was never considered of practical significance in the past, at least until the advent of desktop computers enabled the widespread use of resampling methods. It is the Hodges–Lehmann estimator defined as the median of the $n(n - 1)$ pairwise averages

$$\hat{\Delta} = median_{i \leq j} (X_j + X_i)/2$$

With the aid of the bootstrap, we can determine its precision as easily as that of the mean or the median.

6.2.3. Bootstrap Reproducibility Index

As we saw in the last chapter, a multitude of statistics are available for making comparisons between very large data sets. Ideally, reproducibility of the various methods would be evaluated using independent samples. Although such samples usually do not exist, we can create random samples using the bootstrap.

Ma [2006] made an empirical study of supervised gene screening using the following procedure:

1. Randomly sample $n_1 \sim 0.632 \times n$ subjects from the n observations without replacement.

2. For each bootstrap sample and a fixed gene screening method, compute the marginal statistics T_j^* for gene $j = 1, \ldots, d$. Select the m top ranked genes.

3. Repeat steps 2 and 3 $B = 1000$ times.

4. For gene j, compute O_j: the number of times this gene is included in the m top ranked genes out of the B bootstrap samples.

5. The bootstrap reproducibility index (BRI) for gene j under the chosen screening statistic is defined as $BRI_j = O_j/B$.

6.2.4. Estimation in Regression Models

In general kinetic modeling via positron emission tomography (PET), the _ligand_ concentration at a given brain location at time t, denoted $C(t)$, is given in terms of a convolution:

$$C(t) = g(t; \beta) \otimes C_p(t; \alpha)$$

where $g(t; \beta)$ represents the kinetics of the brain compartmental system and involves unknown rate parameters β. The input function $C_p(t; \alpha)$ represents the concentration of the ligand in the plasma at time t and the input parameters α are estimated separately using data obtained directly from blood samples drawn during the imaging experiment.

The density of the receptor may be estimated by computing a function of the estimated rate parameters. These estimated densities may then be compared among different patient populations or treatment groups. Since there is substantial noise involved with imaging and modeling, in addition to large intrasubject variability, these comparisons often lack power. With standard error estimates, power can be increased dramatically.

The simplest model in common use is a two-compartment model, one compartment for the plasma and one for a brain region of interest (ROI). The $g(t; \beta)$ function is a simple exponential, so the concentration of the ligand in a given brain location is given by

$$C(t) = k_1 e^{-k_2 t} \otimes C_p(t; \alpha)$$

for unknown rate parameters k_1 and k_2. The plasma concentration function is commonly modeled by a piecewise expression:

$$C_p(t) = \alpha_0 I(0 \leq t < t^*) + \sum_{i=1}^{3} \alpha_i \exp[-t\lambda_i] I(t \geq t^*)$$

Plasma data (consisting of measured concentration based on total radioactivity counts corrected for rate of decay and metabolism) are taken over time from arterial samples while the subject is being scanned: $X_i = C_p(s_i) + \varepsilon_i, i = 1, \ldots, n$. The parameters involved in the expression of $C_p(t)$ are estimated using nonlinear least squares.

Through the PET imaging modality, concentration is measured for each of several time-ordered "frames" at each location on a 3D grid. In some applications, aggregate measures are computed for several anatomically defined ROI by combining all data from voxels contained within the ROI. Assuming the two-compartment model, the brain data are given by

$$Y_i = C(t_i) + \varepsilon_i, \quad i = 1, \ldots, n$$

The bootstrap must incorporate the variability of the estimation of the input parameters into the calculation of standard errors for the rate parameters. As suggested by Efron and Tibshirani [1993; Section 9.5] bootstrap samples of the Y data are obtained from the residuals of the model fit rather than from the raw data.

A two-stage bootstrap was employed by Ogden and Tarpey [2006]:

1. The *outer loop*: For each $k = 1, \ldots, B_1$,
    (a) obtain a bootstrap sample of the α data and calculate the resulting $\hat{\alpha}_k$ value.
    (b) The *inner loop*: For each $\ell = 1, \ldots, B_2$, using the computed value of $\hat{\alpha}_k$, obtain a bootstrap sample of the Y data and compute $\hat{\beta}_{\ell(k)}$.
    (c) Compute $\hat{u}_k$, the average of the $\hat{\beta}_{\ell(k)}$ values.
    (d) Compute $\hat{\Gamma}_k$, the sample variance–covariance matrix of the $\hat{\beta}_{\ell(k)}$ values.
2. Estimate $\text{Var}(E[\hat{\beta}|\hat{\alpha}])$ by computing the sample variance–covariance matrix of the $\hat{\mu}_k$ values.
3. Estimate $E[\text{Var}(\hat{\beta}|\hat{\alpha})]$ by computing the matrix average of the $\hat{\Gamma}_k$ values.
4. The bootstrap estimate of the (unconditional) variance–covariance matrix of $\hat{\beta}$ is the sum of these two components.

Note that in the two-stage algorithm, each $\hat{\mu}_k$ is an estimate of the corresponding $E[\hat{\beta}|\hat{\alpha}_k]$ and $\hat{\Gamma}_k$ is an estimate of $\text{Var}(\hat{\beta}|\hat{\alpha}_k)$.

6.3. CONFIDENCE INTERVALS

The problem with single-value (point) estimates is that we will always be in error, unless we can sample the entire population. The solution is an interval estimate or confidence interval where we can have confidence that the true value of the population functional we are attempting to estimate lies between some minimum and some maximum value with a prespecified probability. For example, to obtain a 90% confidence interval for the variance of sixth-grader's heights, we might exclude 5% of the bootstrap values from each end of Figure 6.1. The result is a confidence interval whose lower bound is 52 cm^2 and whose upper bound is 95 cm^2. Note that our original point estimate of 76.7 cm^2 is neither more nor less likely than any other value in the interval [52, 95].

6.3.1. Testing for Equivalence

Often we would like to demonstrate (or are obliged by some regulatory agency to demonstrate) that two methods or two treatments or two devices are equivalent; that is, their expected results are within d of one another.

1. We begin by making J observations by the existing method and K by the new method (we may start out by intending to draw equal-sized samples, but we seldom realize this ambition). Both J and K ought be larger than 25, the larger the better, if this method is to be of value.
2. Draw a bootstrap sample consisting of J observations from the first sample and K from the second. Compute the difference in means (or medians) of the two.
3. Draw 400 such bootstrap sample pairs. If $100 - \alpha$ percent of the differences are smaller than d in absolute value, you will have demonstrated equivalence.

To determine the probability of detecting a nonequivalent treatment, assume that the treatment means differ by $D > d$. Add D to all the observations in the first sample and repeat the bootstrap procedure. Your estimate of this probability (the power of the test) is the percentage of differences that are greater than d in absolute value.

6.3.2. Parametric Bootstrap

If we know the form of the population distribution, we should make use of the information. The *parametric* bootstrap can provide answers even when no textbook formulas exist.

The parametric bootstrap is particularly valuable when confidence intervals are required for statistics such as the variance and the 90th percentile that depend heavily on values in the tails.

Suppose we know the observations come from a normal distribution and want an interval estimate for the 95th percentile. We would draw repeated bootstrap samples from the original sample, use the mean and variance of the bootstrap sample to estimate the parameters of a normal distribution, and then derive a bootstrap estimate of the 95th percentile from tables of a $N(0, 1)$ distribution.

We would follow precisely the same procedure if we knew that the observations came from a Weibull or an exponential distribution. Schall [1995] used the parametric bootstrap to assess bioequivalence using parameter estimates obtained from an initial analysis of variance.

R Code

```
#The following R program fits an exponential
  distribution to the data set A
#Then uses a parametric bootstrap to get a 90%
  confidence interval for the IQR of the
  population from which the data set A was taken.
#n=length(A)
#create a vector in which to store the IQR's
IQR = numeric(1000)
#Set up a loop to generate the 1000 IQR's
for (i in 1:1000) {
bA=sample(A, n, replace=T)
```

```
IQR[i] = qexp(.75,1/mean(bA))-qexp(.25, 1/mean(bA))
}
quantile (IQR, probs = c(.05,.95))
```

6.3.3. Blocked Bootstrap

A glance at a sixth-grade classroom will reveal, much to the dismay of the sixth grade boys, that girls are taller than boys. To construct bootstrap confidence intervals in situations where the effects of age, gender, concurrent medications, and other factors may well overwhelm that of the object of the investigation, separate bootstrap samples should be drawn from each of the factor combinations.

To investigate brain reorganization at the individual level, statistical tests have to deal with the uncertainty related to fMRI time series, which cannot be regarded as independent and identically distributed samples from a given process. The *circular block bootstrap* should be employed in order to respect the temporal dependencies of the data. This consists of drawing blocks of the time series rather than independent observations. The block length needs to be adapted to the range of temporal dependencies and the number of volumes. For adequate block lengths, this method preserves spatial correlation, and formally leads to consistent confidence intervals of spatial correlations; see Lahiri [2003]. For more details on its application, see Bellec, Marrelec, and Benali [2008].

6.3.4. Balanced Bootstrap

The purpose of the balanced bootstrap is to ensure that while an observation may be missing from a particular bootstrap sample or be replicated several times, in the total number B of bootstrap samples, each observation will appear the same number B of times. The result is a slight decrease in the variance of the bootstrap and a reduction in computation time. This reduction is not of practical significance in most applications with the exception of those of high dimensionality such as cluster analysis and correspondence analysis.

To obtain a balanced bootstrap, fill a vector with the observations in the original sample replicated B times. Randomly permute the observations in the vector, then take successive bootstrap samples starting at one end of the permuted vector, with the final bootstrap sample being taken at the opposite end.

The balanced bootstrap is recommended for the analysis of multidimensional contingency tables.

6.3.5. Adjusted Bootstrap

Although complicated to program and compute, the adjusted bootstrap is of value when estimating the prediction error for very large arrays.

Pick J bootstrap learning sets of sizes $l_j n, j = 1, \ldots, J$. Compute the repeated leave-one-out bootstrap estimate with bootstrap learning sets of size $l_j n$ as follows.

For every original sample x, leave out one observation at a time and denote the resulting sets by $x_{(-1)}, \ldots, x_{(-n)}$. From each leave-one-out set $x_{(-i)}$, draw B_1 bootstrap learning sets of size ln. Build a prediction rule on every bootstrap learning set generated from $x_{(-i)}$ and apply the rule on the test observation x_i.

The repeated leave-one-out bootstrap estimate is the misclassification rate calculated across all the bootstrap runs and all n observations. It can be expressed as

$$\hat{e}_n^{RLOOB}(l) = \sum_{i=1}^{n} \sum_{b_i=1}^{B_1} I\{y_i \neq r(t_i, x_{(-i)}^{*,b_i})\}/nB_1$$

where $x_{(-i)}^{*,b_i}$ is the b_ith bootstrap learning set of size ln drawn from the set $x_{(-i)}$.

Let e_{m_j} denote $\hat{e}_n^{RLOOB}(l_j)$ where $m_j = c(l_j)n$ is the expected number of distinct original observations in a bootstrap learning set of size l_jn.

Fit an empirical learning curve of the form $e_{m_j} = am_j^{-c} - b$ with $j = 1, \ldots, J$. The estimates for the parameters are obtained by minimizing the nonlinear least squares function

$$\sum_{j=1}^{J}\{e_{m_j} - am_j^{-c} - b\}^2$$

The adjusted bootstrap estimate for the prediction error is given by $\hat{e}_n = \hat{a}n^{-\hat{c}}$ It is the fitted value on the learning curve as if all original observations contributed to an individual bootstrap learning set.

6.3.6. Which Test?

Giannakakis et al. [2006] used a bootstrap rather than a permutation test to ascertain whether the mean directed transfer function of subjects with specific learning difficulties was significantly lower than that of healthy controls. Unless there was strong reason to believe that the variability among subjects was greater among those with specific learning difficulties, the permutation test ought to have been employed.

As Jiang and Simon [2007] note: "With the large amount of noisy information and limited sample sizes in microarray studies, it is often preferable to provide conservative estimates for the prediction error in order to avoid false positive reports on the prediction models. The ordinary bootstrap, the bootstrap cross-validation and the .632 bootstrap are thus considered less competitive because they provide anti-conservative estimates under some circumstances, even though they can work well in terms of the mean squared errors in strong signal situations."

For small to moderate sized samples, they suggest using the adjusted bootstrap method since it remains conservative, thus avoiding overly optimistic assessment of a prediction model, and does not suffer from extremely large bias or variability in comparison to other methods.

As discussed in Section 3.2.5, Storey et al. [2005] used cubic splines to model time-course data for microarrays. The direct way to deal with the correlation structure of time-course data is to bootstrap from among the entire set of residuals corresponding to an individual. This requires a balanced experiment in which each individual has been sampled at the exact same time points. If the number of subjects is small, bootstrapping at the individual level will not be accurate enough.

To circumvent these difficulties, the residuals can be transformed to be uncorrelated, these uncorrelated residuals bootstrapped, and then transformed back to a set of residuals having the original correlation structure.

The permutation procedure adopted by Sohn et al. [2009] for the analysis of time-course data described in Section 5.3.1 requires a smaller size and is computationally far simpler.

6.4. DETERMINING SAMPLE SIZE

In this section, we make repeated use of the data we have in hand, the so-called empirical distribution, to determine the appropriate sample size for a more detailed analysis.

Most software used for sample size determination is based on the assumption that the data is drawn from a normal distribution. But if you are convinced that your measurements have quite a different distribution, the bootstrap can help you determine the optimal sample size. The procedure is an iterative one.

1. Collect some preliminary data values from the control distribution.

2. Collect some preliminary data values from the population to be tested *or* create an artificial test sample of the sort you expect to encounter. (For example, add random numbers between 1 and 5 to all the values in the test sample.)

3. Choose a test statistic.

4. Estimate the desired sample size S_0 under the assumption that your data is normally distributed (or has a multivariate normal distribution).

5. Draw 1000 bootstrap samples from the control data you've collected and use them to determine the 95th percentile to use as a cutoff value for a test at the 5% level.

6. Draw 1000 bootstrap samples from your test data set and determine how many exceed the cutoff value you determined in step 5. This is the power of your test for samples of size S_0.

7. If the power is too low, increase the sample size and repeat steps 5 and 6.

   If the power is higher than it needs to be, decrease the sample size and repeat steps 5 and 6.

   Or stop as you've determined the optimal sample size.

6.4.1. Establish a Threshold

Sekihara, Sahani, and Nagarajan [2005] also make use of the empirical distribution, bootstrapping from it repeatedly in order to establish a threshold for target activities in magnetoencephalography.

In most studies using positron emission tomography (PET) or functional magnetic resonance imaging (fMRI),[18] the

---

[18]See Chapter 4 for an explanation of these methodologies.

task stimulus generally elicits the target cortical activities as well as other activities associated with the target activities. The control stimulus is designed to elicit only the latter activities. Then, by calculating the statistical difference between the images measured with the two kinds of stimuli, the target activities can be revealed.

The empirical probability distribution of these activities is derived using the time-course reconstruction in the control period. The empirical distribution is then used for deriving an appropriate value for a threshold below which nontarget activities that exist in both the task and control measurements can be eliminated.

6.5. VALIDATION

If choosing the correct functional form of a model in the case of a single variable presents difficulties, consider that in the case of k variables, there are k linear terms (should we use logarithms? should we add polynomial terms?) and $k(k-1)$ first-order cross products of the form $X_i X_k$. Should we include any of the $k(k-1)(k-2)$ second-order cross products?

The obvious solution (similar to belling the cat) is to confine the task to just the "important" variables, the ones that bear a causal relationship to the variable we are trying to predict or the hypothesis we are proposing to test.

Gail Gong [1986] constructed a logistic regression model based on observations Peter Gregory made on 155 chronic hepatitis patients, 33 of whom died. The object of the model was to identify patients at high risk. The 19 potential explanatory variables were derived from medical histories, physical examinations, X-rays, liver function tests, and biopsies.

Gong's logistic regression models were constructed in two stages. At the first stage, each of the explanatory variables was evaluated on a univariate basis. Thirteen of these variables proved significant at the 5% level when applied to the original data. A forward multiple regression was applied to these 13 variables and four were selected for use in the predictor equation.

She then proceeded to take several hundred *bootstrap* samples of the 155 patients. Each bootstrap sample was obtained by drawing with replacement from the 155 vectors of observations. That is, she drew samples of the patients, not of individual values. Thus one patient's observations might be missing completely from the bootstrap sample, while another's might be duplicated two or even three or more times. When one bootstraps, the original sample takes the place of the population, a process that clearly saves both money and time.

Computer programs that perform the bootstrap typically draw a sample with replacement from the one-dimensional list of patient's names, then construct a multidimensional matrix of the associated data. In Gong's case, the size of the bootstrap sample, m, was the same as the original sample size, n. In cases where the number of variables is an order of magnitude greater than the number of observation vectors, it is preferable to take $m < n$, a procedure known as subsampling.

When she took bootstrap samples of the 155 patients, the R^2 values of the final models associated with each bootstrap sample varied widely.[19] Not reported in her article, but far more important, is that while two of the original four predictor variables always appeared in the final model derived from a bootstrap sample of the patients, five other variables were incorporated in only some of the models.

---

[19] R^2 is a measure of the amount of the dependent variable's variability that is explained by the predictors.

Long et al. [2001] used subsamples of their subjects to validate their analysis of a microarray, while Pepe et al. [2003] bootstrapped.

Qiu et al. [2006] used the bootstrap to assess the stability of gene selection in microarray data analysis for several of the univariate statistics and Type I error-controlling procedures reviewed in the last chapter.

Figure 6.2 illustrates the use of the bootstrap to assess the stability of dendrograms developed via cluster analysis.

6.5.1. Cluster Analysis

Cluster analysis differs from *classification*, considered in the next chapter, in that its purpose is to establish relationships or classifications among phenomena, for example, to define market segments, to find structural similarity among chemical compounds, and similarity in function among genes.

Just as the bootstrap may be used to validate regression models, so it is also recommended for validating cluster analysis. Bootstrapping cluster analysis begins with creating a number of simulated data sets based on the statistical model.

If ANOVA is the appropriate model as proposed by Kerr and Churchill [2001], the bootstrap-simulated data sets would take the form

$$y^*_{ijkg} = \hat{\mu} + \hat{A}_i + \hat{D}_j + (\hat{A}D)_{ij} + \hat{G}_g + (\hat{A}G)_{ig} + (\hat{V}G)_{kg}$$
$$+ (\hat{D}G)_{jg} + \varepsilon^*_{ijkg}$$

where a caret ($\wedge$) over a term means the estimate from the original model fit. The error terms are drawn *with replacement* from the studentized residuals of the original model fit. Repeat

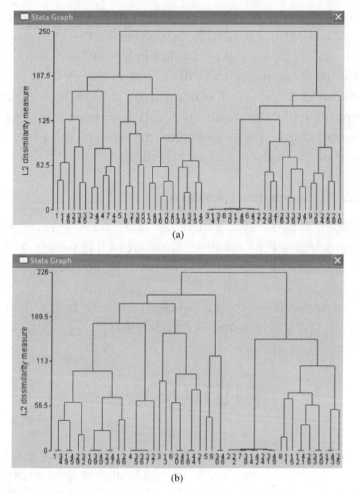

FIGURE 6.2 Dendrograms of (a) the original sample and (b) a bootstrap sample.

the clustering procedure on each simulated data set:

$$y^* \to \hat{r}^* \to \hat{C}^*$$

to obtain a collection of bootstrap clusterings $\{\hat{C}^*\}$.

This approach is inadequate absent some measure of consistency. Dolnicar and Leisch [2010] recommend that one compute a cluster index, the Calinski–Harabas index (Milligan and Cooper [1985]) or the Rand index (Hubert and Arabie [1985]), for each bootstrap data set. As the number of clusters as well as the composition of those clusters will vary from bootstrap sample to bootstrap sample, only clusters that appear with some regularity should be the object of further investigation.

6.5.2. Correspondence Analysis

Consider a two-way contingency table with R rows and C columns and cell probabilities denoted by $\prod_{ij}$ (where $\sum\sum\prod_{ij} = 1$). Let the row and column marginals be denoted by $\prod_{i+}$ and $\prod_{+j}$.

Correspondence analysis is based on the fact that, with appropriate x and y scores and λ correlations, the $\prod_{ij}$ terms can be written

$$\prod_{ij} = \prod_{i+}\prod_{+j}\prod_{.k}(1 + \sum_m \lambda_m x_{im} y_{jm})$$

where $1 \leq m \leq M = 2\ \min(r,c) - 1$ and $\lambda_1 \geq \lambda_2 \geq \cdots \geq \lambda_M$.

Similarly, we may reduce a multicolumn gene expression matrix to its canonical form by placing the data in a three-dimensional matrix whose elements are d_{ijk}, where i represents a specific gene ($i = 1, \ldots, I$), j the time points ($j = 1, \ldots, J$), and $k = 0$ for control and $k = 1$ for disease. The expected values $\prod_{ijk}$ of the d_{ijk} are estimated from the marginals of this matrix.

We write

$$\prod_{ijk} = \prod_{i.}\prod_{j}\prod_{.k}(1 + \sum_m \lambda_m x_{imk} y_{jmk})$$

where $1 \leq m \leq M = 2\min(r,c) - 1$ and $\lambda_1 \geq \lambda_2 \geq \cdots \geq \lambda_M$. In this canonical representation, the departure from independence is partitioned into M components, where the x's and y's are the scalings or score systems for the categories of the gene and time variables and the λ is the canonical correlation between the gene and time variables with the given scalings.

Consider the following hypotheses:

C0 : $\lambda_1 = \lambda_2 = \cdots = \lambda_M = 0$, that is, total independence

C1 : $\lambda_2 = \cdots = \lambda_M = 0$

C2 : $\lambda_3 = \cdots = \lambda_M = 0$

and so forth

Reiczigel [1996] suggests the following testing procedure. First, model C0 is tested and if it fits, the process ends. If it has to be rejected, testing continues with model C1 and so on, until one of the models turns out to be acceptable.

The main steps of testing model Ck, are as follows:

1. Estimate the parameters (λ, x, and y) by least squares;
2. Compute the entries in the contingency table, assuming model Ck is true;
3. Block bootstrap from this table and compute a test statistic t^* for each bootstrap sample;
4. Compare the actual value of t to a desired level critical value of the bootstrap distribution of the t^* values.

Kishino and Waddell [2000] were the first to apply correspondence analysis to microarrays using a parametric approach. The bootstrap adopted here is recommended.

Tan et al. [2004] were the first to propose a bootstrap procedure for inferring the significance of contributions once a time-course experiment is in <u>canonical form</u>. Their approach differs from the method above in that they first compute the test statistics, then reduce the matrix of test statistics to canonical form, and apply the bootstrap to estimates derived from the latter. Unfortunately, we were unable to confirm their method of reduction and none of this article's many authors responded to frequent requests for clarification.

6.6. BUILDING A MODEL

The noisier the data, the greater the proportion of outlying values, the more likely one is to *overfit* a model. That is, in attempting to fit each and every value in the data at hand, the resultant model proves of little predictive value when confronted with new data.

The RANSAC algorithm due to Fischler and Bolles [1981] uses as small an initial data set as feasible, then attempts to enlarge it through the addition of compatible data. Suppose we are given a model that requires a minimum of n data points to determine its free parameters (e.g., $n = 2$ for a straight line, $n = 3$ for a circle), and a sample S of size $N > n$. A typical example is SIFT feature points in a stereo matching problem. We may have $M = 1000$ data items and need only $N = 8$ matched pairs for estimating the fundamental matrix.

Set $k = 1$.

Select a subsample S_k of size n from S at random and use it to derive a model M_1. Add all points of the original sample to the subsample that are within some predetermined error tolerance of M_1. Call the combined subsample S_k^*.

Case 1: Size$(S_k^*)/N$ is greater than some predetermined threshold T. Use the data in S_k^* to determine a new model M and quit.

Case 2: Size$(S_k^*)/N$ is less than some predetermined threshold T. If the number of random subsamples k is less than some predetermined value K, set $k = k + 1$ and repeat the process. Otherwise consider adding an additional parameter to the model.

Alignment of retinal images via an extension of this method due to Yang and Medioni [1992] proves accurate only for small regions. The dual-bootstrap iterative closest point algorithm due to Stewart, Tsai, and Roysam [2003] starts from an initial transformation estimate of the "bootstrap region" of the mapped image and expands it into a globally accurate final transformation estimate. See Figure 6.3.

Rather than using a single, fixed transformation model, different models are used as the bootstrap region expands, starting

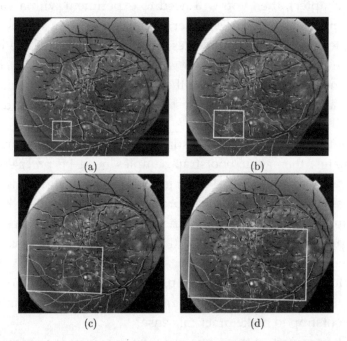

FIGURE 6.3 Illustrations of the dual-bootstrap iterative closest point (ICP) algorithm in retinal image registration. Reproduced from Stewart et al. [2003] with permission. Copyright © 2003 IEEE.

from a simple model for the initial bootstrap region and gradually evolving to a higher order multiparameter model as the bootstrap region grows to cover the entire image overlap region. Model selection techniques are used to automatically select the transformation model for each bootstrap region.

6.7. HOW LARGE SHOULD THE SAMPLES BE?

Should each bootstrap sample have the same number of observations as the original sample? The answer to this question depends on the use to which the bootstrap samples will be put. If your objective is to use the original sample to estimate the size a sample from the same population ought to be to obtain a predetermined power level (a topic we discuss at length in the next chapter), then you will need to experiment with a variety of sample sizes—some smaller, some larger than the original sample. Bootstrap sample sizes less than the size of the original sample are recommended when the original sample size is extremely large, as would be the case when analyzing microarrays, and consistent estimates are desired; see Lee and Pun [2006].

But if your objective is to estimate the precision of the original estimates, then your bootstrap samples need be precisely the same size as the original sample. Here is why: Recall that the precision of a sample mean is proportional to the square root of the number of observations in the sample. Smaller samples yield less precise estimates while estimates based on larger samples are more precise.

How many bootstrap samples should one take? For that matter, how large ought the original sample be for application of the bootstrap to make practical sense?

The answers to these two questions are interdependent. The quickest way to spot that the original sample size is inadequate is to take two sets of 100 bootstrap samples and compare the results. If they are radically different, this means that the original sample lacked sufficient information concerning the estimator in question. Taking 10,000 bootstrap samples as Chernick [2008] suggests is like putting pancake makeup on a zit—it merely conceals and does not eliminate the problem.

Any estimates that depend on the tail of a distribution, such as dispersions, bias, and the 90th and 95th percentiles, require quite large samples to begin with (greater than 1000 observations) in order to be estimated with sufficient precision.

6.8. SUMMARY

In this chapter, we described the bootstrap and several of its variants. You learned how they could be applied to the analysis of very large data sets for the purpose of validation, point and interval estimation, and the determination of sample size and cutoff thresholds. In our final chapter, you'll see how investigators make repeated use of the bootstrap in cross-validating classification methods.

6.9. TO LEARN MORE

The bootstrap has its origins in the seminal work of Jones [1956] and Efron [1979]. Among its earliest applications to real-world data are found in Makinodan et al. [1976] and Gong [1986]. For further examples of the wide applicability of the method, see Chernick [2008]. Among the earliest examples of its application to very large data sets are Lele and Richtsmeier [1995] for X-rays and van der Laan and Bryan [2001] for microarrays. The sample size for the former study was much too small for a bootstrap to be appropriate.

An R package for hierarchical clustering with p-values may be downloaded from http://www.is.titech.ac.jp/~shimo/prog/pvclust/. See Suzuki and Shimodaira [2004].

CHAPTER 7

CLASSIFICATION METHODS

The practical objective of microarrays, EEGs, MRI, and other medical and nonmedical images is to obtain a diagnosis. Is this individual at risk for cancer? Is his/her brain function normal? Is this geographical area arable? Or a likely source of oil?

In this chapter, we consider five methods of classification: nearest neighbor, Bayesian networks, naive Bayes classifier, discriminant analysis, logistic regression, and decision trees. We compare the most commonly used decision tree algorithms. As is done throughout the text, a list of available software is provided.

With all methods, we distinguish between the *training data* used to create the classifying scheme and the *test data* to which we will apply it. Finally, we emphasize the importance of validation with a completely distinct test set.

7.1. NEAREST NEIGHBOR METHODS

To classify a new observation, we locate its k closest neighbors in the training set and establish its classification by majority vote.

Analyzing the Large Numbers of Variables in Biomedical and Satellite Imagery, First Edition.
Phillip I. Good.
© 2011 John Wiley & Sons, Inc. Published 2011 by John Wiley & Sons, Inc.

Following Fix and Hodges [1951], the value k to be employed is determined by <u>cross-validation</u>, that is, by running the nearest neighbor classifier on each observation in the training set. Its distance to all of the other training set observations is computed and it is classified by the nearest neighbor rule. The value of k for which the cross-validation error rate is smallest is retained for future use.

The measure of distance to be employed will depend on the application. One possibility for use with microarrays is the correlation coefficient,

$$r_{\bar{x},\bar{x}'} = \frac{\sum_{j=1}^{p} (x_j - \bar{x})(x_j' - \bar{x}')}{\sqrt{\sum_{j=1}^{p} (x_j - \bar{x})^2} \sqrt{\sum_{j=1}^{p} (x_j' - \bar{x}')^2}}$$

where $\mathbf{x} = (x_1, \ldots, x_p)$ is the gene expression data from one of the samples, and $\mathbf{x}'$ is the gene expression data from a second sample.

Dudoit, Fridlyand, and Speed [2002] report that for the classification of tumors using gene expression data the nearest neighbor method is at least as good as and, in some cases, is superior to the more sophisticated methods discussed in the balance of this chapter.

7.2. DISCRIMINANT ANALYSIS

Fisher linear discriminant analysis (FLDA) is based on finding linear combinations $\mathbf{xa}$ of the gene expression levels $\mathbf{x} = (x_1, \ldots, x_p)$ with large ratios of between-phenotypes to within-phenotype (but among subjects) sum of squares. The prior distribution for the intensity data need not be known.

The *maximum likelihood (ML) discriminant rule* classifies an observation by that which gives it the largest likelihood, that is, by $C(\mathbf{x}) = \arg\max_k pr(\mathbf{x}|y = k)$. In the case when the measured intensities are independent and normally distributed, the discriminant rule is linear and given by

$$C(x) = \arg\min_k \sum_{j=1}^{p} (x_j - \bar{x}_{kj})^2/s_j^2$$

Dudoit, Fridlyand, and Speed [2002] found that for the data they analyzed, ignoring correlations between genes and applying the linear discriminant produced impressively low misclassification rates compared to decision trees.

7.3. LOGISTIC REGRESSION

When our choice is limited to two possible classifications (though both could be composite as in "likely to get breast cancer," and the other "unlikely to get breast cancer"), we may specify the probability that the first diagnosis is correct as $(1 + \exp[-z])^{-1}$, where the expected value of z is

$$\beta_0 + \beta_1 x_1 + \beta_2 x_2 + \beta_3 x_3 + \cdots + \beta_k x_k$$

Confidence intervals for the coefficients β_k are obtained via the bootstrap.

The coefficients of a logistic regression do not lend themselves to a causal interpretation, nor do the higher-order interactions that are conspicuously absent from the preceding formula. The errors (residuals) in a logistic regression are assumed to be independent, identically distributed, and normally distributed. Not one of these three assumptions is likely to be realized in practice.

7.4. PRINCIPAL COMPONENTS

The principal components method makes use of regression after first reducing the number of variables to be considered. Specifically,

1. Compute (univariate) standard regression coefficients for each feature.

2. Form a reduced data matrix consisting of only those features whose univariate coefficient exceeds a threshold estimated by cross-validation in absolute value.

3. Compute the first few principal components of the reduced data matrix.

4. Use these principal components in a regression model to predict the outcome.

The package superPc for R to perform this method may be downloaded from `http://www-stat.stanford.edu/~tibs/` `superpc/tutorial.html`.

7.5. NAIVE BAYES CLASSIFIER

The naive Bayes classifier is a probabilistic algorithm based on the assumption that the attribute values are conditionally independent given the target values. The naive Bayes classifier applies to learning tasks, where each instance x can be described as a vector of attribute values $a_1, a_2, \ldots, a_n$ and the target function $f(x)$ can take on any value from a finite set V. When a new instance x is presented, the naive Bayes classifier assigns to it the most probable target value by applying the rule

$$f(x) = \arg\max_{v_j \in V} P\{v_j\} \prod_i P\{a_i | v_j\}$$

The naive Bayes builds its classifier by estimating the different $P(v_i)$ and $P(a_i | v_j)$ terms based on their frequencies over the training data.

7.6. HEURISTIC METHODS

Three heuristic methods also have been used for classification based on very large data arrays. See Emmert-Steib and Dehmar

[2008] for a description of the Bayesian network approach, Mukherjee [2003] on the use of support vector machines, and Shoemaker and Lin [2005] for the application of neural networks.

We do not discuss these methods here as the result of their application does not lend itself to a cause-and-effect interpretation with the consequence that their user is unable to build on the results in the design of further experiments. The exception, to be found in several of the articles in Emmert-Steib and Dehmar [2008], is when knowledge of the relationships between clinical covariates, genes, and genotypes is incorporated into the network model.

7.7. DECISION TREES

Decision trees have two advantages: They are distribution free and the variables that form the tree may be dependent. The nodes or branch points of the decision trees most commonly encountered in practice bifurcate on the basis of one of two possible criteria:

1. *Categorical*: Subject has a specific property versus Subject does not have a specific property.
2. *Continuous*: X (for subject) $\leq k$ versus X (for subject) $> k$.

Pei [2006] provides just such a decision tree for diagnosis of autosomal dominant polycystic kidney disease.

But with which variable should the tree begin? The classification and regression tree (CART) algorithm orders the predictors in their order of importance should they be used in the absence of other information, and starts the tree with the most important. Within each of the two branches, it then reorders the variables again (variables that are highly correlated with the root variable will no longer appear so important) and uses the new most important variable as the basis of the next node.

Hossain, Hossan, and Bailey [2008] use the area under the ROC curve to determine a node based on its performance in isolation and then use the misclassification rate to choose a split point. The curve is obtained by plotting the curve of the true positive rate (*Sensitivity*) versus the false positive rate (1 − *Specificity*) for a binary classifier by varying the discrimination threshold. Ferri et al. [2002] also make use of the area under the ROC curve.

7.7.1. A Worked-Through Example

The data for this example, which may be downloaded from http://lib.stat.cmu.edu/datasets/boston, consists of the median value (MV) of owner-occupied homes in about 500 U.S. census tracts in the Boston area, along with several potential predictors including:

1. CRIM per capita crime rate by town,
2. ZN proportion of residential land zoned for lots over 25,000 square feet,
3. INDUS proportion of nonretail business acres per town,
4. CHAS Charles River dummy variable (= 1 if tract bounds river; 0 otherwise),
5. NOX nitric oxides concentration (parts per 10 million),
6. RM average number of rooms per dwelling,
7. AGE proportion of owner-occupied units built prior to 1940,
8. DIS weighted distances to five Boston employment centers,
9. RAD index of accessibility to radial highways,
10. TAX full-value property tax rate per $10,000,
11. PT pupil–teacher ratio by town,
12. LSTAT % lower status of the population.

Variable	LSTAT	RM	DIS	NOX	PT	INDUS	TAX	AGE	CRIM	ZN	RAD
Score	100	89.56	28.55	26.21	25.15	22.59	19.88	15.69	13.18	11.46	4.93

FIGURE 7.1 Variable selection by CART using the OLS criteria, in order of importance.

The initial ordering of these variables by CART® software in accordance with the variable's value as discriminators is shown in Figure 7.1.

Using R to develop a decision tree:

```
library("tree")
bos.tr=tree(MV~
CRIM+ZN+INDUS+CHAS+NOX+RM+AGE+DIS+RAD+TAX+PT
 +B+LSTAT)
summary(bos.tr)
```

Variables actually used in tree construction:

```
[1] "RM" "LSTAT" "DIS" "CRIM" "PT"
```

Number of terminal nodes: 9
The command

```
plot(bos.tr); text(bos.tr, srt=90)
```

yields the tree displayed in Figure 7.2. Note that this particular decision tree has gone beyond classification to predict the median value of houses.

We might also have used R and stepwise regression to make the prediction:

```
summary(step(lm(MV~
CRIM+ZN+INDUS+CHAS+NOX+RM+AGE+DIS+RAD+TAX+PT+B+
LSTAT)))
```

yielding the result

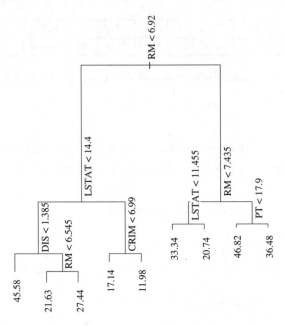

FIGURE 7.2 Regression tree for predicting median home value.

```
MV = 36.3 - 0.11CRIM + 0.05ZN + 2.7CHAS + 17.38NOX
     + 3.80RM + 1.49DIS + 0.30RAD-0.01 TAX + 0.94PT
     + 0.009B - 0.52LSTAT
```

Because of the differences in the units that are employed in measuring each variable, the values of these coefficients do not aid us in discerning which are the most important predictors.

7.8. WHICH ALGORITHM IS BEST FOR YOUR APPLICATION?

Different tree algorithms yield different results. For example, Trujillano et al. [2008], when attempting to classify critically ill patients in terms of their severity, found that CART made use of five variables and eight decision rules, CHAID seven variables and 15 rules, and C4.5 six variables and ten rules.

Classification and regression tree builders (CART® and cTree) make binary splits based on either yes/no or less than/greater than criteria. The choice of variable to be used at each split and the decision criteria are determined by tenfold cross-validation. The resultant tree can be *pruned*; that is, branches below a certain number of levels can be combined. Results can be improved by making use of relative losses and a priori frequencies of occurrence.

The approach used by the CART and cTree algorithms has two drawbacks:

1. *Computational complexity.* An ordered variable with n distinct values at a node induces $(n - 1)$ binary splits. For categorical variables, the order of computations increases exponentially with the number of categories; it is $(2^{M-1}-1)$ for a variable with M values.

2. *Bias in variable selection.* Unrestrained search tends to select variables that have more splits. As Quinlan and Cameron-Jones (1995) observe: "for any collection of training data, there are 'fluke' theories that fit the data well but have low predictive accuracy. When a very large number of hypotheses is explored, the probability of encountering such a fluke increases."

One alternative is the FACT algorithm. Instead of combining the problem of variable selection with that of split criteria, FACT deals with them separately. At each node, an analysis of variance F-statistic is calculated for each ordered variable. The variable with the largest F-statistic is selected and linear discriminant analysis is applied to it to find the cutoff value for the split.

Categorical variables are handled less efficiently by transforming them into ordered variables. If there are J classes among the data in a node, this method splits the node into J subnodes. See Figure 7.3 and compare with Figure 7.4.

QUEST (quick, unbiased, efficient, statistical tree), like CART, yields binary splits and includes pruning as an option, but offers

```
Read 150 cases (5 attributes) from iris.data
Decision tree:
petal length <= 1.9: setosa (50)
petal length > 1.9:
:...petal width < 1.7: virginica (46/1)
    petal width >= 1.7:
    :...petal length <= 4.9: versicolor (48/1)
  petal length > 4.9: virginica (6/2)
```

FIGURE 7.3 Fisher iris data as fit by C5.0. The observations may be downloaded from http://archive.ics.uci.edu/ml/machine-learning-databases/iris/.

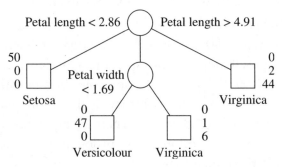

FIGURE 7.4 Iris data using the FACT method. The triple beside each terminal node gives the number of cases in the node of Setosa, Versicolour, and Virginica, respectively.

the further advantage that it has negligible variable selection bias and retains the computational simplicity of FACT. QUEST can easily handle categorical predictor variables with many categories. It uses imputation instead of surrogate splits to deal with missing values. If there are no missing values in the data, QUEST can optionally use the CART algorithm to produce a tree with univariate splits. See Figure 7.5.

Warning: While CART, FACT, and QUEST yield different error rates for the IRIS data, this may not be true for all data sets.

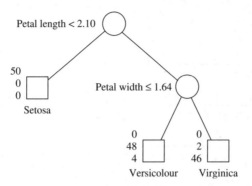

FIGURE 7.5 Iris data using the QUEST method. The triple beside each terminal node gives the number of cases in the node of Setosa, Versicolour, and Virginica, respectively.

CHAID (chi-squared automatic interaction detector) divides the population into two *or more* groups based on the categories of the "best" predictor of a dependent variable. It merges values that are judged to be statistically homogeneous (similar) with respect to the target variable and maintains all other values that are heterogeneous (dissimilar). Each of these groups is then divided into smaller subgroups based on the best available predictor at each level. The splitting process continues recursively until no more statistically significant predictors can be found (or until some other stopping rule is met).

7.8.1. Some Further Comparisons

When Trujillano et al. [2008] used decision trees to classify the severity of critically ill patients, the decision tree they generated based on the CHAID methodology used seven variables and began with the variable INOT. Fifteen decision rules were generated with an assignment rank of probability ranging from 0.7% to a maximum of 86.4%. Certain variable values, Glasgow value, age, and $(A-a)O_2$, were divided into intervals with more than two possibilities. See Figure 7.6.

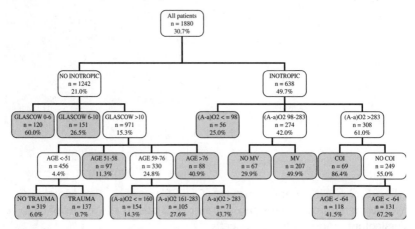

FIGURE 7.6 Classification tree by CHAID algorithm. (Note that some splits give rise to three or four branches.) The gray squares denote terminal prognostic subgroups. INOT: inotropic therapy; (A-a)O2 gradient: alveolar–arterial oxygen gradient (mmHg); MV: mechanical ventilation; COI: chronic organ insufficiency.

The decision tree generated by Trujillano et al. [2008] based on the CART methodology used only five variables and also began with INOT. It generated eight decision rules with an assignment rank of probability ranging from 5.9% to a maximum of 71.3%.

Their C4.5 model used six variables (the five common variables plus mean arterial pressure) and generated ten decision rules. The probabilities ranged between 7.6% and 76.2%. In contrast to the other decision trees, the first variable was the point value on the Glasgow scale.

7.8.2. Validation Versus Cross-validation

While cross-validation may be used effectively in developing a decision tree, to compare decision tree algorithms with one another or with heuristic methods such as neural networks or support vector machines, the training set and the test set must be completely distinct. Moreover, the same training set and data set must be used for all methods to be compared.

7.9. IMPROVING DIAGNOSTIC EFFECTIVENESS

The diagnostic effectiveness of decision trees can be increased in several ways:

1. Specifying the proportions of the various classifications to be expected in the population to be tested.
2. Specifying the relative losses to be expected if observations are misclassified. The default is to treat all misclassifications as if they had equal impact. Following a mammogram with a more accurate ultrasound is expensive, but not when compared with the cost of allowing a malignancy to grow out of control.
3. Boosting.
4. Bagging and other ensemble methods.

7.9.1. Boosting

C5, a much enhanced version of C4.5, provides the ability to improve on the original decision tree via boosting. Instead of drawing a succession of independent bootstrap samples from the original instances, boosting maintains a weight for each instance—the higher the weight, the more the instance influences the next classifier generation as first suggested by Quinlan [1996]. Initially, all weights are set equally, but at each trial, the vector of weights is adjusted to reflect the performance of the corresponding classifier, with the result that the weight of misclassified instances is increased so that the weak learner is forced to focus on the hard instances in the training set.

7.9.2. Ensemble Methods

Ensemble methods combine multiple classifiers (models) built on a set of resampled training data sets, or generated from various classification methods on a training data set. This set of classifiers form a decision committee, which classifies future coming samples. The classification of the committee can be by simple vote or by weighted vote of individual classifiers in the committee.

The essence of ensemble methods is to create diversified classifiers in the decision committee. Aggregating decisions from diversified classifiers is an effective way to reduce bias existing in individual trees. However, if classifiers in the committee are not unique, the committee has to be very large to create diversity in the committee.

Bagging is an older bootstrap ensemble method that creates individuals for its ensemble by training each classifier on a set that is a random sample with replacement from the original.

To compare the performance of several machine training algorithms based on decision trees, Ge and Wong [2008] conducted a series of statistical analyses on mass spectral data obtained from premalignant pancreatic cancer research. They found that the classifier ensemble techniques outperformed their single algorithm counterparts in terms of the consistency in identifying the cancer biomarkers and the accuracy in the classification of the data.

7.9.3. Random Forests

A random forest (RF) is a classifier consisting of a collection of individual tree classifiers. Three steps are employed during the learning phase.

1. Select *ntree*, the number of trees to grow, and *mtry*, a number no larger than the number of variables; the typical default is mtyy = sqrt[ntree].
2. For $i = 1$ to *ntree*:

- Draw a bootstrap sample from the training set. Call those not in the bootstrap sample the "out-of-bag" data.
- Grow a "random" tree, where, at each node, the best split is chosen among *mtry* randomly selected variables (i.e., a bootstrap sample). The tree is grown to maximum size and not pruned back.
- Use the tree to predict out-of-bag data.
3. In the end, use the predictions on out-of-bag data to form majority votes.

Prediction of test data is done by majority vote of the predictions from the ensemble of trees.

In addition to the out-of-bag error rate, RFs also provide a proximity measure and an importance measure. To obtain the proximity measure, the entire tree is examined. If two observations reside in the same terminal node, then the proximity between them is increased by 1. The normalized proximity measure is the total count divided by the total number of trees. An outlier measure for the jth observation can be defined as $1/S_k(\text{proximity}[j,k])^2$, where a large value (say, greater than 10) indicates a possible outlier that may be neglected.

To obtain the importance value for a gene G in a particular pathway, Pang, Kim, and Zhao [2008] permute the expression values for this gene among the out-of-bag observations. The permuted expression values are then used to get classifications. A large reduction in performance accuracy would be expected for an informative gene. Margin is defined as percentage of votes for the correct class minus arg $\max_c$ (percentage of votes for another class c). The mean decrease in accuracy for gene G is the reduction of the margin across all individuals divided by the number of cases when gene G is randomly permuted.

7.10. SOFTWARE FOR DECISION TREES

Unfortunately, the software for analyzing decision trees resolves into two categories—the expensive but comprehensive, and the free but lacking in features.

Pricey but Complete. (Includes adjustments for relative losses and population proportions, boosting, and bagging.)

C5.0. A scaled-down version may be downloaded without charge from `http://www.rulequest.com/download.html`.

CART. A trial version may be obtained from `http://salford-systems.com`.

CHAID is part of IBM SPSS AnswerTree. See `http://www.spss.com/software/statistics/answertree/`.

Free but Lacking in Features. Adaboost. See `http://www2.fml.tuebingen.mpg.de/raetsch/downloads/software/adaboost-reg-rbf`.

CRUISE may be downloaded without charge from `http://www.stat.wisc.edu/~loh/cruise.html`.

C4.5 (an older version of C5 lacking many of its features) may be downloaded without charge from `http://www2.cs.uregina.ca/~hamilton/courses/831/notes/ml/dtrees/c4.5/tutorial.html`.

Excel users can form decision trees by downloading Ctree, a macro-filled Excel spreadsheet, from `http://www.geocities.com/adotsaha/CTree/CtreeinExcel.html`.

QUEST may be downloaded without charge from `http://www.stat.wisc.edu/~loh/quest.html`.

The rpart package may be downloaded from `http://cran.r-project.org/web/packages/rpart/index.html`.

7.11. SUMMARY

We have considered a number of methods for the classification of subjects on the basis of their biomedical data, but have focused on the use of decision trees.

Decision trees should be used for classification and prediction rather than regression whenever the following is true:

- Predictors are interdependent and their interaction often leads to reinforcing synergistic effects.
- A mixture of continuous and categorical variables, highly skewed data, and large numbers of missing observations add to the complexity of the analysis.
- Data sets may have errors or missing values.

Decision tree algorithms vary in the results they yield and the supporting software varies widely in its capabilities. As we shall see in the next chapter, preprocessing and variable reduction methods can greatly increase their efficiency.

While tenfold cross-validation or the bootstrap may be used to develop a classifier, validation via a separate data set distinct from the original data set is necessary to assess the result or to compare it with other classifiers.

CHAPTER 8

APPLYING DECISION TREES

The assignment of classifications to very large biological and medical data sets require three essential steps:

1. Standardization of the images so all have a common orientation and scale.
2. Reduction of variables.
3. Classification.

Which methods are used in steps 1 and 2 depend strongly on the nature of the data. In this chapter, we consider the methods that have been employed when classifying galactic images, sonographs, MRI, EEGs, MEG, mass spectral data, and microarrays.

8.1. PHOTOGRAPHS

For the analysis of photographic images, the three stages are:

Analyzing the Large Numbers of Variables in Biomedical and Satellite Imagery, First Edition.
Phillip I. Good.
© 2011 John Wiley & Sons, Inc. Published 2011 by John Wiley & Sons, Inc.

- Standardization: images are rotated, centered, and cropped, creating images invariant to color, position, orientation, and size, in a fully automatic manner.
- Principal component analysis reduces the dimensionality of the data.
- Machine learning via a decision tree.

To standardize the images of galaxies, de la Calleja and Fuentes [2004] first derive the galaxy's main axis, which is given by the first eigenvector (the eigenvector with the largest corresponding eigenvalue) of the covariance matrix. They rotate the image so that the main axis is horizontal. They crop it by eliminating columns that contain only background pixels that fall below a predetermined intensity threshold. Finally, all images are stretched to a standard size as shown in Figure 8.1.

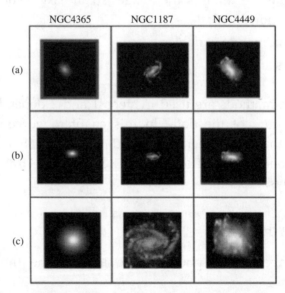

FIGURE 8.1 (a) Original images, (b) rotated images, and (c) cropped images of galaxies NGC4365, NGC1187, and NGC4449. Reproduced with permission from Springer.

Principal component analysis transforms a number of (possibly) correlated variables into a smaller number of uncorrelated variables called principal components (PCs). de la Calleja and Fuentes used a subset of the components corresponding to 75%, 80%, and 85% of the information in the data set, respectively.

They employed four distinct machine learning methods:

- Naive Bayes classifier
- C4.5 decision tree algorithm
- Random forest
- Bagging

Unfortunately, they did not compare the methods by applying the predictors they derived to test sets that were distinct from the training set, so that a direct comparison among methods is not possible.

8.2. ULTRASOUND

If a suspicious lump in the breast is detected by a routine mammogram, an ultrasound is the immediate follow-up. Multiple sonographic findings are necessary to evaluate solid breast masses, even by an experienced interpreter, including shadowing, single nodule, spiculation, angular margins, thick echogenic halo, microlobulation, taller-than-wide, hypoechogenicity, calcifications, and duct extension or branch pattern.

Kuo et al. [2002] made use of texture analysis and decision trees to classify breast tumors studied with different ultrasound units.

To evaluate texture parameters, they made use of the covariance defined as the expected value of $[g(i,j) - \zeta]$ $[g(i + \Delta i, j + \Delta j) - \zeta]$, where $g(i,j)$ is the intensity value at the position (i,j), $(\Delta i, \Delta j)$ is the distance between two pixels, and ζ is the gray level average of the contrast region (about $5 \times 5 = 25$ pixels).

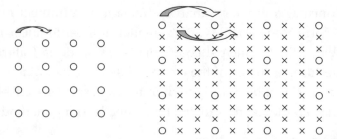

FIGURE 8.2 Diagram illustrating the resolution adjustment relationship between two images with different resolutions. Reproduced with permission from Elsevier Ltd.

They used the C5.0 algorithm to generate a decision tree. The covariance texture parameters were used as the inputs to construct the decision tree model and for further diagnosis.

The 243 images obtained with the SDD 1200 system were used as the training set; while 259 images obtained with the HDI 3000 system composed the testing set. As each ultrasound system has its own resolution and contrast (see Figure 8.2), Kuo et al. [2002] found a preliminary resolution adjustment was necessary, the details of which are provided in their article. After adjustment, their decision tree was able to classify images in the test set with an accuracy of 90%, sensitivity 95%, specificity 85%, positive predictive value 86%, and negative predictive value 94%. See Figure 8.2.

8.3. MRI IMAGES

The texture features of MRI images also lend themselves to diagnosis as they are readily calculated, correlated to tumor pathology, robust to changes in MRI acquisition settings, invariant to changes in image resolution, and unaffected by the corruption of the MRI image by magnetic field inhomogeneity.

The training set employed by Juntu et al. [2010] consisted of T_1-weighted MRI images from 135 patients. The set was artificially inflated by selecting multiple regions from a single patient (albeit the regions were not selected from the same MRI image but selected from several MRI images).

For feature computation, 681 tumor subimages of size 50×50 pixels were imported to the software package MaZda 3.20. This program provides for computation of 300 texture features based on statistical, wavelet filtering, and model-based methods. To ensure the consistency of the calculated texture feature across all the tumor subimages, they wrote some macros for the MaZda program that read tumor subimages and calculated the tumors' texture features with the same texture analysis parameters setting.

The authors made use of three classifiers: neural networks, support vector machines, and the C4.5 decision tree. A tenfold cross-validation procedure was used to train the classifiers. As can be seen from Figure 8.3, the neural network is inferior to both

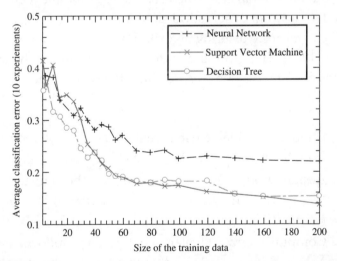

FIGURE 8.3 The learning curves were produced by training the three classifiers with different training data sizes. Each point in each learning curve is the averaged classification error of ten random cross-validation experiments. Reproduced from Juntu et al. [2010] with permission from John Wiley & Sons.

the other methods, while the decision tree approach is preferable for the smaller sample sizes one is likely to encounter in practice.

MaZda tutorials and software for use under MS Windows may be downloaded from `http://www.eletel.p.lodz.pl/programy/mazda/`.

8.4. EEGs AND EMGs

When Günes, Polat, and Yosunkaya [2010] omitted the variable reduction stage, the results were far from satisfactory. Following extraction of features from all-night polysomnogram recordings that included both EEG (electroencephalogram) and chin EMG (electromyogram) signals, classification of the sleep stages via a decision tree had an accuracy of only 42%. The accuracy using only the EEG signal was 38%. But when K-means clustering-based feature weighting was employed, classification accuracy using the EEG signal alone increased to 92% and the inclusion of EMG data was not of statistical significance.

The K-means clustering algorithm was developed by J. Mac-Queen [1967]. The grouping is done by minimizing the sum of squares of distances between data and the corresponding cluster centroid.

1. Choose K initial cluster centers $z_1, z_2, \ldots, z_K$ randomly from the n points $\{X_1, X_2, \ldots, X_n\}$.

2. Assign point $X_i, i = 1, 2, \ldots, n$ to the cluster $C_j, j \in \{1, 2, \ldots, K\}$ if $|X_i - z_j| < |X_i - z_p|, p = 1, 2, \ldots, K$ and $j \leq p$.

3. Compute new cluster centers as follows $z_i^{new} = \sum_{X_j \in C_i} X_j / n_i, i = 1, 2, \ldots, K$, where n_i is the number of elements belonging to the cluster C_i.

4. If $|z_i^{new} - z_i| < \varepsilon$ for $i = 1, 2, \ldots, K$, then terminate. Otherwise continue from step 2.

The ratios of means of features to their centers are calculated and these ratios are multiplied by the data point of
each feature before being processed by the C4.5 decision tree
algorithm.

8.5. MISCLASSIFICATION COSTS

Some misclassifications are more serious than others. By varying
the costs assigned to false negatives (Cfn) and false positives
(Cfp) as shown in Figure 8.4, the sensitivity and specificity of
an assay can be adjusted to the required needs. For example,
Ladanyi et al. [2004] designated false negatives as three times
more costly than false positives. In their work, they also made
use of boosting, tenfold cross-validation, and an independent
external data set for validation.

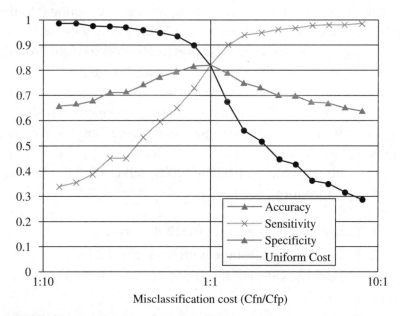

FIGURE 8.4 Reproduced from Ladanyi et al. [2004] with permission
from John Wiley & Sons.

8.6. RECEIVER OPERATING CHARACTERISTIC

Receiver operating characteristic (ROC) analysis is based on plotting the true positive rate (TPR) on the y-axis and the false positive rate (FPR) on the x-axis. Each classifier (a gene in the case of a microarray) is represented by a single point in the ROC diagram. The ROC curve is used to evaluate the discriminative performance of binary classifiers. This is obtained by plotting the curve of the true positive rate (*Sensitivity*) versus the false positive rate (1 − *Specificity*) for a binary classifier by varying the discrimination threshold.

Parodi, Izzotti, and Muselli [2005] propose that at the first step, the combination of a few genes with high areas under the ROC curve (AUC) be identified to allow the best separation between the two groups. In the next steps, the same procedure is repeated recursively on the remaining genes to generate a panel of classification rules. For each class, a score proportional to the AUC is then assigned to each sample. The scores generated at each step are summed with the previous scores to yield, in each rule, the genes with the highest AUC. (See Figure 8.5.)

A drawback of their method, particularly with noisy data sets, is that it is based on the choice of very high specificity thresholds, which in some circumstances may leave some samples in a "not decision" subspace.

Hossain, Hossan, and Bailey [2008] advocate the following four-step procedure:

1. Calculate the area under the ROC curve.
2. Determine the attribute with the highest AUC.
3. Choose the splitting threshold based on the misclassification rate.
4. Stop growing the tree for a node when the AUC for an attribute is ≥ 0.95 or ≤ 0.5 to avoid overfitting.

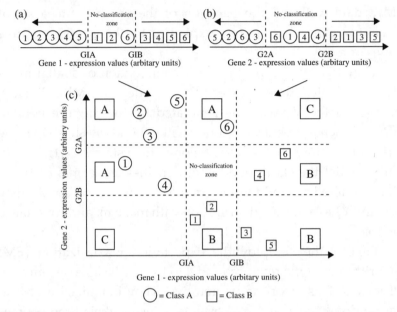

FIGURE 8.5 Method for the identification of classification rules in microarray data. (**a**, **b**) For each selected gene expression, the two cutoff values that allow the classification of the highest number of samples without error (highest specificity) are identified (e.g., G1A and G1B for gene 1 and G2A and G2B for gene 2, respectively). (**c**) Two (or more) genes are then combined to reach the best classification rate. In cross-validation, samples falling in "A" areas are assigned to the class A, whereas samples falling in "B" areas are assigned to the other class. Samples falling in "C" areas are assigned to the class receiving the highest score, based on the two AUC (as illustrated in the text). Reproduced from Parodi, Izzotti, and Muselli [2005] with permission from *Oxford Journals*.

To obtain a copy of their software, write to `hossain@csse.unimelb.edu.au`.

8.7. WHEN THE CATEGORIES ARE AS YET UNDEFINED

8.7.1. Unsupervised Principal Components Applied to fMRI

Golland, Golland, and Malach [2007] simultaneously estimate the optimal representative time courses that summarize the fMRI data well and the partition of the volume into a set of disjoint regions that are best explained by these representative time courses.

They treat the data as a linear combination of spatial maps with associated time courses. They model the fMRI signal $\mathbf{Y}$ at each of the V voxels as generated by the mixture $p_Y\{y\} = \sum_{s=1}^{N_s} \lambda_s p_{Y|S}\{y|s\}$ over N_s conditional likelihoods $p_{Y|S}\{y|s\}$. λ_s is the prior probability that a voxel belongs to system $s \in \{1, \ldots, N_s\}$. They model the class-conditional densities as normal distributions centered around the system mean time course; that is, they assume $\mathbf{Y}$ has a normal distribution with mean $\mathbf{m}_s$ and covariance matrix $\sum_s$.

Finally, they employ the expectation maximization (EM) algorithm to fit the mixture model to the fMRI signals from a set of voxels. This algorithm is iterative. They initialized each run by randomly selecting N_s voxels and using their time courses as an initial guess for the cluster means. They assumed the covariances were all zero and initialized the variances to the data variance across voxels. They initialized the prior probabilities to a uniform distribution over K clusters, and experimented with various values for K.

At the nth iteration, they computed the $(n + 1)$st values as follows:

$$p^n\{s|y_v\} = \frac{\lambda_s^n N(y_v; m_s^n, \sum_s^n)}{\sum_r \lambda_r^n N(y_v; m_r^n, \sum_r^n)}$$

$$\lambda_s^{n+1} = \sum_v p^n\{s|y_v\}/V$$

The estimated mean values are

$$m_s^{n+1} = \frac{\sum_v y_v p^n\{s|y_v\}}{\sum_v p^n\{s|y_v\}}$$

and the diagonal elements of the covariance matrix are

$$\sum_{s}^{n+1}(t,t) = \frac{\sum_v (y_v(t) - m_s^{n+1}(t))^2 p^n\{s|y_v\}}{\sum_v p^n\{s \mid y_v\}}$$

where $p^n\{s|y_v\}$ is the estimate of the posterior probability that voxel v belongs to system s. MatLab code for an EM algorithm for clustering may be downloaded from `http://www.mathworks.nl/matlabcentral/fileexchange/3713`.

To summarize the results of the segmentation across subjects, they employ an approximate algorithm that matches the label assignments in pairs of subjects with the goal of maximizing the number of voxels with the same label in both subjects until convergence. In practice, this algorithm quickly (1–2 passes over all subjects) finds the correct label permutation in each subject. The proportion of voxels that achieved perfect agreement across subjects was statistically significant.

8.7.2. Supervised Principal Components Applied to Microarrays

Using supervised principal components, Bair and Tibshirani [2004] develop diagnostic procedures to predict the survival of future patients based on both their gene expression profiles and the survival times of previous patients. Their procedure lends itself to situations in which the possible classifications are not known in advance.

Patients with a similar prognosis may respond very differently to the same treatment. One possible explanation is that their seemingly similar problems are actually completely different diseases at the molecular level. Bair and Tibshirani combine gene expression data with clinical data to identify previously undefined cancer subtypes.

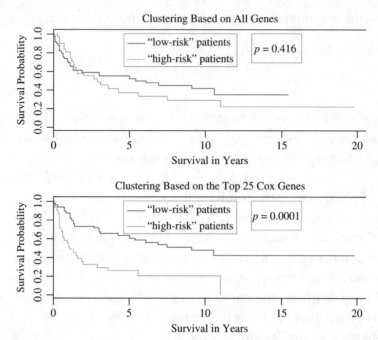

FIGURE 8.6   Comparison of the survival curves resulting from applying two different clustering methods. Reproduced from Bair and Tibshirani [2004] with permission.

First, they ranked all 7399 genes based on their univariate Cox proportional hazards scores derived from 160 patients. They used the 25 highest ranked genes to form clusters in an independent test set derived from 80 patients. The results are shown in Figure 8.6. Conventional clustering techniques applied to the same data failed to identify subgroups that differed with respect to their survival times.

They also calculated the principal components of the training data using only the data for the 17 genes with Cox scores of 2.39 or greater. Following Alter et al. [2000], they used the first few columns of V as continuous predictors of survival for each patient where $V = X^{\mathrm{T}}UD^{-1}$, and the element x_{ij} of X denotes the expression level of the ith gene in the jth patient.

To find a predictor based on the first k principal components, perform the following procedure:

- Fit a Cox proportional hazards model using the first k columns of V for the *training set* as predictors.
- Calculate the matrix V^* replacing the training data with the test data.
- Take a linear combination of the first k columns of V^* using the Cox regression coefficients obtained for the trainings set and use the resulting sum as a continuous predictor of survival.

8.8. ENSEMBLE METHODS

To compare the performance of several machine learning algorithms based on decision trees, Ge and Wong [2008] conducted a series of statistical analyses on mass spectral data obtained from premalignant pancreatic cancer research. They found that the classifier ensemble techniques outperformed their single algorithm counterparts in terms of their consistency in identifying the cancer biomarkers and their accuracy in the classification of the data.

Rank is determined by the probability of the two means of the disease and control groups in the training set being significantly different. Most frequent features are determined by the frequency with which each feature appears in the top 10 list in these ten runs and are ranked by their frequency. See Figure 8.7.

8.9. MAXIMALLY DIVERSIFIED MULTIPLE TREES

Bagging is an older bootstrap ensemble method that creates individuals for its ensemble by training each classifier on a set that is a random sample with replacement from the original. A quicker and more efficient way to create diversity is to include

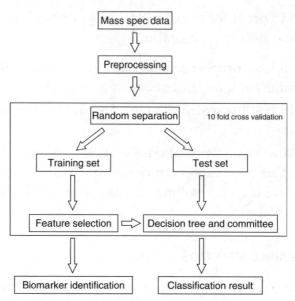

FIGURE 8.7 Workflow used in the analysis of spectral data by Ge and Wong [2008]. Reproduced with permission.

only unique trees in the ensemble. Hu et al. [2006] recommend the following procedure:

1. All samples with all genes are used to build the first decision tree.

2. After the decision tree is built, the used genes are removed from the data.

3. All samples with remaining genes are used to build the next decision tree. Again, the used genes are removed.

This process repeats until the number of trees reaches the preset number. As a result, all trees are unique and do *not* share common genes.

The MDMT method of Hu et al. [2006] has two stages.

First, training in accordance with the three steps described above:

```
train(D, T, n)
```

```
INPUT: A Microarray data set D, and the
     number of trees n.
OUTPUT: A set of disjoint trees T
let T = 0
for i = 0 to n-1
call C4.5 or C5 to build tree t on D
remove genes used in t from D
T = c(T,t)
end for
Output T
```

Note that a gene containing noise or missing values only affects one tree but not multiple trees.

In the second and final stage, the predicted class of a coming unseen sample is determined by the weighted votes from all trees. The value of each vote is weighted by accuracy of the associated tree in making predictions. The majority vote is endorsed as the final predicted class.

```
CLASSIFY(T, x, n)
INPUT: A set of trained trees T, a test
     sample x, and the number of trees n.
OUTPUT: A class label of x
let vote = 0
for j = 1 to n
let class be the class outputted by T[j]
vote(class) = vote(class) + accuracy(T[j])
end for
Output c that maximizes vote(c)
```

Their software may be downloaded from http://www.cis.unisa.edu.au/~lijy/MDMT/MDMT.html.

8.10. PUTTING IT ALL TOGETHER

To compare next generation sequencing with microarrays, Stiglic, Bajgot, and Kokol [2010] used the bootstrap to create subsets, ranking methods as well as percentage-of-gene-overlap score, gene set enrichment employing permutations and meta analysis, and decision trees.

Initially, the bootstrap was used to create k subsets of original data sets. Eleven gene ranking methods from the Bioconductor package Gene-Selector were used to construct the original list of ranked genes l_0 and k ranked lists $l_1, l_2, \ldots, l_k$ of genes on each subset. Average POG score, *POGavg* was measured by averaging all pairwise comparisons of k ranked gene lists with the original ranking on the initial data set.

An analysis of gene set enrichment using the GSEA application was performed on each of the eight data sets. Gene set permutation was used instead of phenotype permutation due to the small sample size. See Figure 8.8.

Their meta-learning-based GSEA automates comparisons of multiple GSEA results that otherwise would have to be done manually. See Figure 8.9.

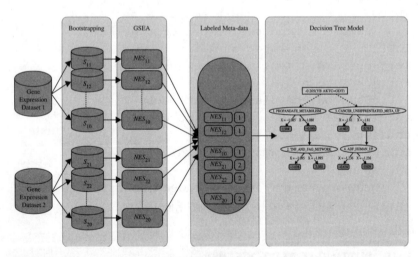

FIGURE 8.8 Workflow of the GSE-MLA procedure describing the process from the initial data sets (e.g., next generation sequencing vs. microarrays) to the final decision tree model. Reproduced from Stiglic, Bajgot, and Kokol [2010] with permission.

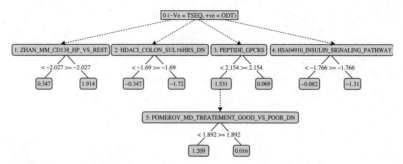

FIGURE 8.9   Representation of GSE-MLA results using ADTree. ADTree explaining the significant differences in gene set enrichment between ODT and TSEQ sample preparation. Reproduced from Stiglic, Bajgot, and Kokol [2010] with permission.

8.11. SUMMARY

The efficiency of a decision tree can be increased by the following:

- Reducing the number of variables to be considered
- Increasing the size of the training set
- Correcting for relative costs
- Correcting for the expected proportions in the target population

8.12. TO LEARN MORE

Schetinin and Schult [2004] combine decision trees and neural networks to classify EEGs. Boulesteix et al. [2008] evaluated a number of microarray-based classifiers.

For more on the KM algorithm, see McLachlan and Krishnan [1997]. For more on *k*-means clustering, see http://www.inference.phy.cam.ac.uk/mackay/itprnn/ps/284.292.pdf.

Lin, Rogers, and Hsu [2001] shift the focus to the localization of disease genes to small chromosomal regions, even at the stage of preliminary genome screening. Eisen et al. [1998] make use of hierarchical clustering to display genome-wide expression patterns.

GLOSSARY OF BIOMEDICAL TERMINOLOGY

Base Pair

In two-stranded DNA, the nucleotide bases are paired with one base on one strand and the complementary base on the other. Adenine forms a base pair with thymine, as does guanine with cytosine.

Cerebral Cortex

A sheet of neural tissue outermost to the cerebrum of the mammalian brain that plays a key role in memory, attention, and perceptual awareness.

DNA and RNA

The DNA (deoxyribonucleic acid) in the nuclei of our cells communicates with the sites of protein synthesis in our cells' cytoplasm via mRNA (messenger ribonucleic acid). Portions of the intertwined double strands of our DNA separate and the

Analyzing the Large Numbers of Variables in Biomedical and Satellite Imagery, First Edition. Phillip I. Good.
© 2011 John Wiley & Sons, Inc. Published 2011 by John Wiley & Sons, Inc.

exposed portion acts as a template for the formation of the corresponding single-stranded mRNA. Two other types of RNA are involved in protein synthesis—transfer RNA (tRNA) and ribosomal RNA (rRNA).

Gene and Genome

A single gene determines the production of a single cell protein. The genome embraces the entirety of genes.

Gene Expression

Most genes remain unexpressed except for brief periods in the life of the cells in which they are located. They become active or expressed when they are needed to provide for the synthesis of specific proteins.

Genotype and Phenotype

The two-stranded nuclear DNA of animals (including us humans) is composed of one strand from the mother and one from the father. Although each strand has the same number of genes in the same positions, the alleles (i.e., the gene variants) on the two strands may or may not be the same. For example, the strand that was contributed by my mother has the alleles for O− blood, while that from my father provides for A+ blood. My phenotype, that is, the outward expression of these genes, is A+.

Grey and White Matter

The brain and spinal cord are composed of grey matter, which contains neural cell bodies, in contrast to white matter, which does not and mostly contains myelinated axon tracts. The color difference arises mainly from the whiteness of myelin. In living tissue, grey matter actually has a grey-brown color, which comes from capillary blood vessels and neuronal cell bodies.

Hybridization

DNA sequencing-by-hybridization is an indirect method for determining whether a gene has been expressed. Two steps are required:

- The target DNA sequence is brought in contact with a microarray chip of short-length nucleotide sequences (probes), so that the whole subset of probes binding to the target, called its sequence spectrum, is determined by some biomedical measurements.
- A combinatorial algorithm is applied to the spectrum to rebuild the sequence.

Indel

A mutation resulting in a colocalized insertion and deletion and a net gain or loss in nucleotides.

Ligand

In the neuroimaging application described in the text, a radioactive biochemical substance (the ligand) is infused into the bloodstream, where it binds to neuroreceptors. The decay of the radioactive isotope in the ligand is measured via positron emission tomography.

Osseous

Relating to bones

Probe

A molecular probe is a group of atoms or molecules that are either radioactive or fluoresce, are attached to other molecules or cellular structures, and are used in studying the properties of these molecules and structures.

Sagittal, Coronal, and Transverse Planes

The standard chest X-ray is taken with one's chest is pressed against the screen, thus enabling a photograph of the coronal plane. If asked to stand sidewise to the screen, the X-ray would be of the sagittal plane. A transverse plane would cut a mermaid so that the human part was on one side and the fish on the other.

Voxel and Pixel

A voxel (volumetric pixel) is a volume element, representing a value on a regular grid in three-dimensional space. A pixel is an area element, representing a value on a regular grid in two-dimensional space.

GLOSSARY OF STATISTICAL TERMINOLOGY

Accuracy and Precision

An *accurate* estimate is close to the estimated quantity. A *precise* interval estimate is a narrow interval. Precise measurements made with a dozen or more decimal places may still not be accurate.

Bootstrap

When we bootstrap, a sample from a population takes on the role of the population. Bootstrap samples are taken with replacement from the original sample. Large random samples will closely resemble the population. Small samples and nonrandom samples may be quite different in composition from the population.

Canonical Form

In the context in which it is used by Tan et al. [2004], canonical form refers to the representation of a matrix of observations as

Analyzing the Large Numbers of Variables in Biomedical and Satellite Imagery, First Edition.
Phillip I. Good.
© 2011 John Wiley & Sons, Inc. Published 2011 by John Wiley & Sons, Inc.

the product of three matrices $U\Lambda V'$, where all the off-diagonal elements of Λ are zero. See, also, Principal Component Analysis.

Curve Fitting

Fitting a curve to a series of points can be accomplished by a variety of methods, the simplest of which, as well as the most common, is linear regression. The term linear is somewhat of a misnomer, as we may not be fitting a line to the data, but rather a polynomial, the regression coefficients being the coefficients of the polynomial. For example,

$$y = a + bt + ct^2 + dt^3 + et^4 + ft^5$$

represents an attempt to fit a fifth-order polynomial to time-course data. The fit would be a perfect one if we have exactly six time points to include. But the resulting curve would include a series of oscillations that may not be appropriate to the phenomena we are attempting to model.

We can avoid this by using a more complex method in which we fit the curve in a piecewise fashion with a series of lower order (most often cubic) polynomials called splines.

Whichever method of curve fitting is used, an essential caveat is that goodness of fit is not prediction.

Deterministic and Stochastic

A phenomenon is *deterministic* when its outcome is inevitable and all observations will take specific values.[20] A phenomenon is *stochastic* when its outcome may take different values in accordance with some probability distribution.

---

[20]These observations may be subject to measurement error.

Dichotomous, Categorical, Ordinal, Metric Data

Dichotomous or binary data have two values and take the form "yes or no," "got better or got worse."

Categorical data have two or more categories such as yes, no, and undecided. Categorical data may be ordered (opposed, indifferent, in favor) or unordered (dichotomous, categorical, ordinal, metric).

Preferences can be placed on an ordered or *ordinal* scale such as strongly opposed, opposed, indifferent, in favor, strongly in favor.

Metric data can be placed on a scale that permits meaningful subtraction; for example, while "in favor" minus "indifferent" may not be meaningful, 35.6 pounds minus 30.2 pounds is.

Metric data can be grouped so as to evaluate it by statistical methods applicable to categorical or ordinal data. But to do so would be to throw away information and reduce the power of any tests and the precision of any estimates.

Distribution, Cumulative Distribution, Empirical Distribution, Limiting Distribution

Suppose we were able to examine all the items in a population and record a value for each one to obtain a *distribution* of values. The *cumulative distribution function* of the population $F[x]$ denotes the probability that an item selected at random from this population will have a value less than or equal to $x : 0 \le F[x] \le 1$. Also, if $x < y$, then $F[x] \le F[y]$.

The *empirical distribution*, usually represented in the form of a cumulative frequency polygon or a bar plot, is the distribution of values observed in a sample taken from a population. If $F_n[x]$ denotes the cumulative distribution of observations in a sample of size n, then as the size of the sample increases $F_n[x] \to F[x]$.

The *limiting distribution* for very large samples of a sample statistic such as the mean or the number of events in a large number of very small intervals often tends to a distribution of known form such as the Gaussian for the mean or the Poisson for the number of events.

Be wary of choosing a statistical procedure that is optimal only for a limiting distribution and not when applied to a small sample. For small samples, the empirical distribution may be a better guide.

Exact Test

The significance level of an exact test is exactly as claimed. Permutation tests are always exact if the observations are exchangeable. The significance levels of many parametric tests are only correct with very large samples.

Exchangeable Observations

Observations are said to be exchangeable if their joint probability distribution is independent of the order in which the observations appear. Independent, identically distributed observations are exchangeable. So, too, are certain dependent observations such as those that result from drawing from an urn or other finite population.

F-Statistic

The F-statistic is the ratio of the between-samples variation to the within-samples variation.

Hotelling's T^2

See Multivariate and Univariate Statistics. Formal definition is provided on page 13.

Hypothesis, Null Hypothesis, Alternative

The dictionary definition of a *hypothesis* is a proposition, or set of propositions, put forth as an explanation for certain phenomena.

For statisticians, a *simple hypothesis* would be that the distribution from which an observation is drawn takes a specific form. For example, $F[x] = N(0, 1)$. In the majority of cases, a statistical hypothesis will be *compound* rather than simple; for example, the distribution from which an observation is drawn has a mean of zero.

Often, it is more convenient to test a *null hypothesis*, for example, that there is no or null difference between the parameters of two populations.

There is no point in performing an experiment or conducting a survey unless one also has one or more *alternate hypotheses* in mind. If the alternative is one-sided, for example, the difference is positive rather than zero, then the corresponding test will be one-sided. If the alternative is two-sided, for example, the difference is not zero, then the corresponding test will be two-sided.

Imputation Versus Surrogate Splits

In determining whether to send a case with a missing value for the best split left or right, the CART decision tree algorithm uses surrogate splits. It calculates to what extent alternative splits resemble the best split in terms of the number of cases that they send the same way. This resemblance is calculated on the cases with both the best split and alternative split observed. Any observation with a missing value for the best split is then classified using the most resembling surrogate split, or if that value is missing also, the second most resembling surrogate split, and so on.

Multiple imputation is a simulation-based approach where a number of complete data sets are created by filling in alternative values for the missing data. The completed data sets may subsequently be analyzed using standard complete-data methods, after which the results of the individual analyses are combined in the appropriate way. The advantage, compared to using missing-data procedures tailored to a particular algorithm, is that one set of imputations can be used for many different analyses.

Kolmogorov–Smirnoff Statistic

This metric determines the distance between two empirical frequency distributions F and G, or an empirical frequency distribution F and a theoretical frequency distribution G, as the largest value of the absolute difference $|F[x] - G[x]|$.

Loss Function

Hypothesis testing is merely another aid to decision making. As one would hope to teach teenagers, decisions have consequences. Bring an unsafe drug or automobile to market and innocent people will suffer and lawsuits may destroy the pharmaceutical firm or automobile manufacturer. Fail to market a safe and effective drug and, again, innocent people will suffer if an equivalent cure is not available. The costs or expected losses of decisions need be taken into account.

In making decisions, statisticians typically make use of one of four possible loss functions:

1. Absolute error, where losses are proportional to the size of the error.
2. Squared error, where large inaccuracies result in disproportionately large losses.
3. Minimum error, where small inaccuracies are treated as irrelevant.
4. Maximum error, when only the worst-case scenario is considered.

Mean, Median, and Expected Value

The mean is the arithmetic average of the observations in a sample or an actual population. The median is the halfway point in a sample or an actual population; that is, half the observations are larger than the median and half are smaller. The expected value is the arithmetic average of the individuals in a hypothetical population (e.g., the expected decrease in blood pressure of all those who have or might someday take a specific dose of a drug). Only in symmetric populations are the population median and the expected value equal.

Metric

To distinguish between close and distant, we first need a metric, and then a range of typical values for that metric. A metric m defined on two points x and y, has the following properties:

$$m(x,y) \geq 0$$
$$m(x,x) = 0$$
$$m(x,y) \leq m(x,z) + m(z,y)$$

These properties are possessed by the common ways in which we measure distance. The third property of a metric, for example, simply restates that the shortest distance between two points is a straight line.

A good example is the standard Euclidian metric used to measure the distance between two points x and y whose coordinates in three dimensions are (x_1, x_2, x_3) and (y_1, y_2, y_3):

$$\sqrt{(x_1 - y_1)^2 + (x_2 - y_2)^2 + (x_3 - y_3)^2}$$

This metric can be applied even when x_i is not a coordinate in space but the value of some variable like blood pressure or laminar flow or return on equity. Hotelling's T^2 is an example of such a metric.

Multivariate and Univariate Statistics

A univariate statistic like Student's t is a function of a single variable. A multivariate statistic like Hotelling's T^2 is a function of more than one variable.

Normal and Nonnormal Distributions

When an observation is the sum of a large number of factors, each of which makes a very small contribution to the total, it will have the normal or Gaussian distribution. The mean, median, and mode of a population of normally distributed random variables are identical. The distribution is symmetric about its mean. Highly skewed distributions like the exponential, and distributions of counts like the Poisson and binomial are distinctly nonnormal. Many of the continuous distributions (e.g., of measurements) that are encountered in practice are a mixture of normal distributions.

Overfitting a Model

When first year college students study Hooke's Law in a physics laboratory, the results of their experiments seldom lie on a straight line, as this law would predict. By overfitting, and adding additional parameters, a curve could be found that would pass through all their points. But this curve would be of little or no value for predictive purposes, nor would it provide the insight into cause and effect that Hooke's Law provides.

Parameter and Statistic

A parameter is some characteristic of a population, for example, the arithmetic mean of the heights of all individuals in the population. A statistic is some function of the observations in a sample, for example, the arithmetic mean of the heights of individuals in a sample.

Parametric Methods and Resampling Methods

To achieve a desired significance level or a desired degree of accuracy, parametric methods require that the observations come from a specific distribution. Resampling methods can make use of information about the distribution of the observations if this information is available, but do not require it.

Principal Component Analysis

The data matrix is put into canonical form and the principal components or eigenvalues selected as described in Section 6.5.2.

Resampling Methods, Nonparametric Tests, and Rank Tests

The resampling methods include both parametric and non-parametric methods. Among the latter are the nonparametric bootstraps, decision trees, permutation tests that use the original observations, permutation tests that make use of the ranks of the original observations (rank tests), and several heuristic methods including neural networks and support vector machines.

Residuals and Errors

A residual is the difference between a fitted value and what was actually observed. An error is the difference between what is predicted based on a model and what is actually observed.

Sample and Population

A population may be real (everyone residing in the city of Los Angeles on May 1, 2010) or hypothetical (all patients who might someday ingest Motrin). A sample is a subset of a population. A random sample from a population is a sample chosen so that each member of the currently available population has an equal chance of being included in the sample.

Significance Level and p-Value

The *significance level* is the probability of making a Type I error. It is a characteristic of a statistical procedure.

The *p-value* is a random variable that depends on both the sample and the statistical procedure that is used to analyze the sample.

If one repeatedly applies a statistical procedure at a specific significance level to distinct samples taken from the same population when the hypothesis is true and all assumptions are satisfied, then the p-value will be less than or equal to the significance level with the frequency given by the significance level.

Splines

See Curve Fitting

Studentization, Student's t

When a statistic is recentered (so that it has a known expected value under the null hypothesis, usually zero) and rescaled (so that it is dimensionless and, with luck, has unit variance) we call the process Studentization. One example is Student's t-statistic for comparing two samples. Its numerator is the mean of one sample minus the mean of the other; its denominator is an estimate of the standard error of the numerator. Variants of Student's t-statistic—Welch's t and the Bayes t—have been similarly Studentized. The only differences among these statistics lies in the nature of the estimate of the standard error.

Type I and Type II Error

A Type I error is the probability of rejecting the hypothesis when it is true. A Type II error is the probability of accepting the hypothesis when an alternative hypothesis is true. Thus a Type II error depends on the alternative.

Type II Error and Power

The power of a test for a given alternative hypothesis is the probability of rejecting the original hypothesis when the alternative is true. A Type II error is made when the original hypothesis is accepted even though the alternative is true. Thus power is one minus the probability of making a Type II error.

Validation Versus Cross-validation

Goodness of fit is not the same as prediction. One can validate a model by attempting to apply it to a set of data completely distinct from the data that was used to develop it. With the various methods of cross-validation described in Chapter 6, models are developed repeatedly on a portion of the data and then applied to the remaining portion. Different methodologies only can be compared via validation, that is, by using one data set to develop the competing models, and a second to test them.

APPENDIX: AN R PRIMER

This primer will help you take full advantage of the large number of programs written in R for use in the analysis of very large data sets. You'll learn to download and install R software, to create, merge, and save data sets in R readable format, and to modify existing R programs to meet your own specific needs.

R1. GETTING STARTED

Your first step should be to download the R package without charge from the website http://cran.r-project.org/. Note that there are Windows, Macintosh, and Unix specific versions. Be sure to make use of this program while you are going through this primer.

R is an *interpreter*. This means as you enter the lines of a typical program, you learn line-by-line whether the command you've just entered makes sense (to the computer) and are able to correct the line if you (or I) have made a typing error.

Analyzing the Large Numbers of Variables in Biomedical and Satellite Imagery, First Edition.
Phillip I. Good.
© 2011 John Wiley & Sons, Inc. Published 2011 by John Wiley & Sons, Inc.

When we run R, what we see on the screen is an arrowhead >. If we type 2 + 3 after the arrowhead and then press the enter key, we see [1] 5.

R reports numeric results in the form of a vector. In this example, the first and only element in the answer vector [1] takes the value 5. If you'd wanted to save the result for later use, you might have written

```
result = 2 + 3
```

To enter the observations 1.2, 2.3, 4.0, 3, and 5.1 for later use, type

```
ourdata = c(1.2, 2.3, 4.0, 3, 5.1)
```

where c( ) is R's concatenate function.

You can easily modify all the observations with a single command by writing, for example,

```
.> newdata=2*ourdata+6
```

If you've never used a programming language before, let me warn you that R is very inflexible. It won't understand (or, worse, may misinterpret) both of the following:

```
ourdata = c(1.2 2.3 4.0 3 5.1)
ourdata = (1.2, 2.3, 4.0, 3, 5.1)
```

If you make a mistake and write

```
2*Ourdata+6 or 2*ourdat + 6 or 2*ourdata + six
```

you also will get an error message.

Hint: If, like me, you make mistakes constantly, you can use the up arrow key ↑ to retrieve the last command; then simply correct the error in typing.

Beware, what works in MSWord does not always work in R. For example, `name = "phil"` will not be understood; `name = "phil"` will.

As we gather more observations, we'll want to update our data vector. Suppose the newly gathered observations are 3.2, 7, and −2. We could write

```
ourdata = c(ourdata, 3.2, 7, -2)
```

or we could first put all the new observations in a new vector and then concatenate the two vectors

```
newdata = c(3.2, 7, -2)
ourdata = c(ourdata, newdata)
```

R1.1. R Functions

R includes many built-in functions. Let's give a few of them a workout. Enter the following data set:

```
classdata = c(141, 156.5, 162, 159, 157, 143.5, 154,
       158, 140, 142, 150, 148.5, 138.5, 161, 153,
       145, 147, 158.5, 160.5, 167.5, 155, 137)
```

`sort(classdata)` to locate the minimum and maximum of this data set, then compute its mean(), median(), var(), standard deviation sqrt(var()), and standard error of the mean. To obtain this last, you'll need to use `length(classdata)`, which gives the sample size.

Want to know exactly what var() does? Or see if there is a function std( )? Pull down the R help menu, then select `Rfunctions (text)` as in Figure R1. Type in the name of the function. Omit parentheses. Following the description of the function, you usually will find several helpful examples.

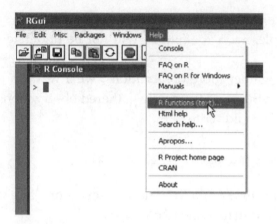

FIGURE R1   Getting help with R functions.

R1.2. Vector Arithmetic

Much of the power of R lies in its ability to do vector arithmetic:
You can speed up your entry of the distance formula

$$S = \sqrt{(x_1 - x_2)^2 + (y_1 - y_2)^2 + (z_1 - z_2)^2}$$

by doing a vector subtraction.

```
> One = c(1,2,3)
> Two = c(3.4, 5.2, 7.1)
> Three = One - Two
> Three
[1] -2.4 -3.2 -4.1
```

R lets you concatenate functions and operations, as in

```
> sqrt(sum((One - Two)**2))
```

R2. STORE AND RETRIEVE DATA

R2.1. Storing and Retrieving Files from Within R

If your data was stored using the R save( ) command:

- Create a directory for your R data files, for example, **`dir.create("C:/Rdata")`**.
- Save your files: **`save(classdata, ourdata, file = "C:/Rdata/pak1.dat")`**.
- To retrieve your data, **`load("c:/rdata/pak1.dat")`**.

Note that our quote marks are straight up and down and that our diagonal slashes are in the opposite direction of those used in MSWindows.

R2.2. The Tabular Format

Let us add additional variables such as age and gender to our class data using the `data.frame()` and `edit()` commands (see Figures R2 and R3):

```
classdata=data.frame(classdata)
class=edit(classdata)
```

Save the data to disk

```
write.tableclass, file="c:/rdata/class.dat")
```

Note that `class.dat` has the following form:

1. The first line of `class.dat` has a name for each variable in the data set.
2. Each additional line or record in `class.dat` has as its first item a row label followed by the values for each variable in the record, separated by blanks.

We can bring `class.dat` into our computer's memory and place it for use in an R dataframe with the command

```
> classdat = read.table("c:/rdata/class.dat")
```

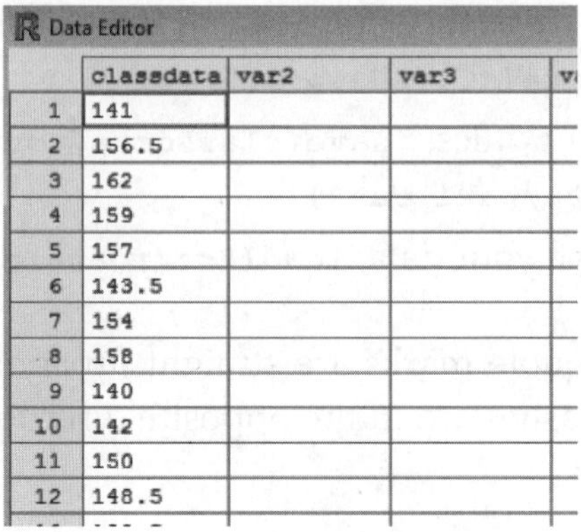

Data Editor

	classdata	var2	var3	v
1	141			
2	156.5			
3	162			
4	159			
5	157			
6	143.5			
7	154			
8	158			
9	140			
10	142			
11	150			
12	148.5			

FIGURE R2 Inside the data editor.

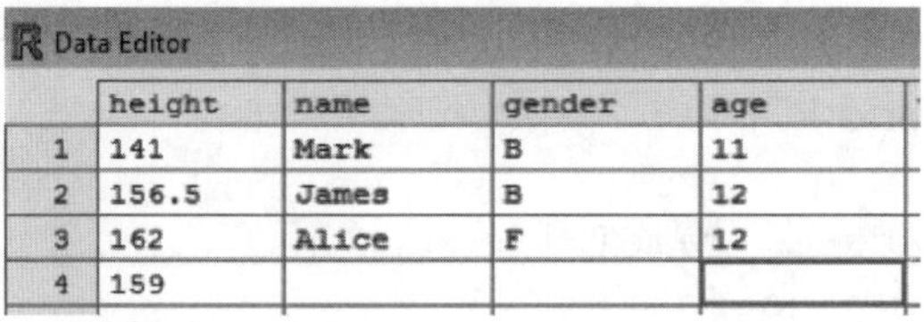

Data Editor

	height	name	gender	age
1	141	Mark	B	11
2	156.5	James	B	12
3	162	Alice	F	12
4	159			

FIGURE R3 After adding variable names and data.

R2.3. Comma Separated Format

Most statistical software as well as Excel can export data files in a comma separated format. Read the file in with a command similar to the following

```
classdat = read.table("c:/mywork/class.dat",
    sep=",")
```

One word of caution: If you are saving column labels (and Excel may do this on its own), you need to use the command

```
classdat=read.table ("c:/mywork/class.dat",
    sep=",", header=T)
```

R3. RESAMPLING

Selecting a test sample with R is simplicity itself; the hard part is figuring out how large a sample to take. Since ours is a hypothetical example, let's generate some fake data with the R command

```
data =1:1000
```

and then select a hundred observations:

```
testsamp = sample (data, 100).
```

One bootstrap sample doesn't tell us very much. A hundred would be better. We'll need to use a **for()** loop.

```
   for (i in 1:100){
 + bootdata= sample (classdata, length(classdata),
     replace=T)
 + median (bootdata)
 + var (bootdata)
 }
```

R3.1. The While Command

Using a **for()** loop makes sense if you know in advance how often a command will be executed. But often you only want to keep repeating a set of instructions until some set of conditions is satisfied. For example, the R command

```
while ( X>0 & Y!= 7){ }
```

translates as "while X is greater than 0 and Y is not equal to 7, do all the instructions between curly brackets { }." You also can make use of symbols such as <=, > =, or | (| means logical "or" and & means logical "and"). Please remember that Y=7 means

FIGURE R4   Menu of R packages.

store the value 7 in the location labeled Y, while `Y==7` is a logical expression that is true only if Y has the value 7. If your R program is giving you wrong answers, it may because you wrote = when you intended `==` .

To work through an ordered vector until some condition is reached, consider the following program:

```
sort(X)
while (x[j] >7) {
        . . . .
     j =j+1
     }
```

R4. EXPANDING R'S CAPABILITIES

R4.1. Downloading Libraries of R Functions

Pulling down on the Package Menu, and then selecting "Load package..." reveals an extensive list of packages (see Figure R4) of additional R functions that you can download and make use of. After you have installed the package, "boot,"for example, it will still be necessary to load it into the computer's memory each time you wish to make use of some program by using the command `library(boot)`.

Not all packages are on this list. For example, to download relaimpo, which provides several metrics for assessing relative importance in linear models, two steps are required. First, go to `http://cran.r-project.org/web/packages/relaimpo/index.html` and download the binary file specific to your type of computer. Then, enter the command in R,

```
utils:::menuInstallLocal(),
```

and you'll be asked to provide the location of the file where you saved the binary code.

To make use of relaimpo at any time in the future, you need to first enter the R command `library(relaimpo)`.

R4.2. Programming Your Own Functions

In how may different ways can we divide eight subjects into two equal–sized groups? Let's begin by programming the factorial function in R.

```
fact  = function (num) {
+ prod(1:num)
+ }
```

Here, num is a parameter of the function **fact()** that will be entered each time we make use of the function, 1:num yields a sequence of numbers from 1 to num and **prod()** gives the product of these numbers.

Or, making use of our knowledge that $N! = N(N-1)!$, we can program in *recursive* form as follows:

```
fact  = function (num) {
    + if (num==0) return (1)
    + else return (num*fact(num-1))
    + }
```

Oops, if num <0, this program will calculate forever. Let's insert a solution message:

```
fact  = function (num) {
    if (num <0) return ("Argument must be positive
    or zero.")
    else {
     if (num==0) return (1)
     else return (num*fact(num-1))
  }
 }
```

Note that the number of closing brackets } *must* equal the number of opening brackets {.

To execute this function, for example, to calculate 7!, type fact(7).

Recently, I wanted to test the robustness of Student's *t* for data that came from nonnormal distributions, so I developed the following program, which makes use of two functions: one that I developed and one that is already provided by R. It's worth reading through the program listing as you'll encounter several new and useful commands along the way. Note my frequent use of comments (preceded by an #) so that a year from now, I'll remember what this program is supposed to do.

```
sim1K= function(N=100,size=c(5,5), alpha=0.1){
#all function parameters are given default val-
ues which may be overridden
#N is the number of repetitions of the simulation
#size is a vector containing the sizes of the
     two samples
#alpha is the significance level of the hypothe-
ses tests I am evaluating
  cnt=0
  for (i in 1:N){
    #generate an artificial data set
    data=gen(size)
    #subdivide this data set into two separate
      samples
    samp1=data[1:size[1]]
    samp2=data[-c(1:size[1])]
    #make use of the p-value from the t.test output
    if(t.test(samp1, samp2,"l")[[3]]<=alpha)cnt
    =cnt+1
    }
  return (cnt)
}
```

Here's what happens if you try to run this program in the computer as is by typing

```
>sim1K()
Error: couldn't find function "gen"
```

which tells us that gen() does not yet exist. So, let's create it.

```
gen = function (size){
     #generate a mixture of normals
     data1=rnorm(size[1],1.5*rbinom(size[1],1,0.3))
     data2=rnorm(size[2],1.5*rbinom(size[2],1,0.7))
   return (c(data1,data2))
}
```

Now we can type and run sim1K().

BIBLIOGRAPHY

Ai-Jun Y, Xin-Yuan S. Bayesian variable selection for disease classification using gene expression data. *Bioinformatics*. 2010;26:215–222.

Alter O, Brown PO, Botstein D. Singular value decomposition for genome-wide expression data processing and modeling. *Proc Natl Acad Sci USA*. 2000; 1997:10101–10106.

Arndt S, Cizadlo T, Andreasen NC, Heckel D, Gold S, Oleary DS. Tests for comparing images based on randomization and permutation methods. *J Cereb Blood Flow Metab*. 1996;16:1271–1279.

Bair E, Tibshirani R. Semi-supervised methods to predict patient survival from gene expression data. 2004. *PLoS Biol*. 2(4): E108.

Bellec P, Marrelec G, Benali H. A bootstrap test to investigate changes in brain connectivity for functional MRI. *Statistica Sinica*. 2008;18:1253–1268.

Belmonte M, Yurgelun-Todd D. Permutation testing made practical for functional magnetic resonance image analysis. *IEEE Trans Med Imaging*. 2001;20:243–248.

Benjamini Y, Yekutieli D. The control of the false discovery rate in multiple testing under dependency. *Ann Statist*. 2001;29:1165–1188.

Benjamini Y, Hochberg Y. Controlling the false discovery rate: a practical and powerful approach to multiple testing. *J R Statist Soc B*. 1995;57:289–300.

Blair RC, Karniski W. An alternative method for significance testing of waveform difference potentials. *Psychophysiology*. 1992;30: 518—524.

Analyzing the Large Numbers of Variables in Biomedical and Satellite Imagery, First Edition. Phillip I. Good.
© 2011 John Wiley & Sons, Inc. Published 2011 by John Wiley & Sons, Inc.

Boulesteix A-L, Strobl C, Augustin T, Daumer M. Evaluating microarray-based classifiers: an overview. *Cancer Inform*. 2008;6:77–97.

Bourgon R, Mancera E, Brozzi A, Steinmetz LM, Huber W. Array-based genotyping in *S.cerevisiae* using semi-supervised clustering. *Bioinformatics*. 2009;25:1056–1062.

Breitling R, Armengaud P, Amtmann A, Herzyk P. Rank products: a simple, yet powerful, new method to detect differentially regulated genes in replicated microarray experiments, *FEBS Lett*. 2004;573:83–92.

Celebi ME, Aslandogan YA. Content-based image retrieval incorporating models of human perception. *Proceedings of the IEEE International Conference on Information Technology: Coding and Computing*. 2004;2:241–245.

Chen Y, Dougherty ER, Bittner M. Ratio-based decisions and the quantitative analysis of cDNA microarray images. *J Biomed Opt*. 1997;2:364–374.

Chernick MR. *Bootstrap Methods: A Practitioner's Guide*, 2nd ed. Hoboken, NJ: Wiley; 2008.

Czwan E, Brors B, Kipling D. Modelling *p*-value distributions to improve theme-driven survival analysis of cancer transcriptome datasets. *BMC Bioinformatics*. 2010;11: 19.

de la Calleja J, Fuentes O. Automated classification of galaxy images. In: *Knowledge-Based Intelligent Information and Engineering Systems*. New York: Springer; 2004;3215: 411–418.

Dobbin K, Simon R. Sample size determination in microarray experiments for class comparison and prognostic classification. *Biostatistics*. 2005;6:27–38.

Dolnicar S, Leisch F. Evaluation of structure and reproducibility of cluster solutions using the bootstrap. *Marketing Lett*. 2010;21:83–101.

Draghici S, Khatri P, Bhavsar P, Shah A, Krawetz SA, Tainsky MA. Onto-Tools, the toolkit of the modern biologist: Onto-Express, Onto-Compare, Onto-Design and Onto-Translate. *Nucleic Acids Res*. 2003;31:3775–3781.

Drigalenko EI, Elston RC. False discoveries in genome scanning. *Genet Epidemiol*. 1997;14:779–784.

Dudoit S, Fridlyand J, Speed TP. Comparison of discrimination methods for the classification of tumors using gene expression data. *JASA*. 2002;97:77–87.

Efron B. Bootstrap methods: another look at the jackknife. *Ann Statist*. 1979;7:1–26.

Efron B. Large-scale simultaneous hypothesis testing: the choice of a null hypothesis. *JASA*. 2004;99.

Efron B, Tibshirani R. *An Introduction to the Bootstrap*. New York: Chapman and Hall; 1993.

Eisen MB, Spellman PT, Brown PO, Botstein D. Cluster analysis and display of genome-wide expression patterns. *PNAS*. 1998;95:14863–14868.

Emmert-Steib F, Dehmar M. *Analysis of Microarray Data: A Network-Based Approach*. Hoboken, NJ: Wiley-VCH; 2008.

Ferri C, Flach P, Hernández-Orallo J. Learning decision trees using the area under the ROC curve. In: *Proc. ICML 2002*. San Francisco: Morgan Kaufmann; 2002.

Fischler MA, Bolles RC. Random sample consensus: a paradigm for model fitting with applications to image analysis and automated cartography. *Commun ACM*. 1981;24:381–395.

Fix E, Hodges J. Discriminatory analysis, nonparametric discrimination: consistency properties. Technical Report 4, Project Number 21-49-004, USAF School of Aviation Medicine, Randolph Field, TX. 1951.

Galán L, Biscay R, Rodriguez JL, Pérez-Ávalo MC, Rodriguez R. Testing topographic differences between event related brain potentials by using non-parametric combinations of permutation tests. *Electroencephalogr Clin Neurophysiol*. 1997;102:240–247.

Gavrilov Y, Benjamini Y, Sarkar SK. An adaptive step-down procedure with proven FDR control under independence. *Ann Statist*. 2009;37:619–629

Ge G, Wong GW. Classification of premalignant pancreatic cancer mass-spectrometry data using decision tree ensembles. *BMC Bioinformatics*. 2008;9:275.

Giannakakis G, Stoitsis J, Trichopoulos G, Nikita K, Papageorgiou Ch, Anagnostopoulos D, Rabavilas A, Soldatos C. Investigation of specific learning difficulties on the information flow in multichannel EEG signals. *5th European Symposium on Biomedical Engineering*, Patras, 7–9 July 2006.

Goeman JJ, Bühlmann P. Analyzing gene expression data in terms of gene sets: methodological issues. *Bioinformatics*. 2007;23:980–987.

Golland P, Golland Y, Malach R. Detection of spatial activation patterns as unsupervised segmentation of fMRI data. *Proc. MICCAI 2007*, Part I, 110–118. New York: Springer-Verlag; 2007.

Gong G. Cross-validation, the jackknife and the bootstrap: excess error in forward logistic regression. *JASA*. 1986;81:108–113.

Good PI. *Introduction to Statistics Via Resampling Methods and R*. Hoboken, NJ: Wiley; 2005.

Good PI. *Resampling Methods*. Basel: Birkhauser; 2006.

Good PI, Hardin J. *Common Errors in Statistics*, 3rd ed. Hoboken, NJ: Wiley; 2008.

Good PI, Lunneborg L. Limitations of the analysis of variance. The one-way design. *J Modern Appl Statist Methods*. 2006;5:41–43.

Gresham D, Douglas M, Ruderfer DM, Pratt SC, Schacherer J, Dunham MJ, Botstein D, Kruglyak L. Genome-wide detection of polymorphisms at nucleotide resolution with a single dna microarray. *Science*. 2006;311:1932–1936.

Günes S, Polat K, Yosunkaya S. Efficient sleep stage recognition system based on EEG signal using *k*-means clustering based feature weighting. *Expert Systems with Applications*. 2010;37:7922–7928.

Harmony T, Fernández T, Fernández-Bouzas A, Silva-Pereyra J, Bosch J, Diaz-Comas L, Galán L. EEG changes during word and figure categorization. *Clin Neurophysiol*. 2001;112:1486–1498.

Hayden D, Lazar P, Schoenfeld D. Assessing statistical significance in microarray experiments using the distance between microarrays. *PLoS One* 4(6): e5838. doi:10.1371/journal.pone.0005838.

Hossain MM, Hossan MR, Bailey J. ROC-tree: a novel decision tree induction algorithm based on receiver operating characteristics to classify gene expression data. *Proceedings of the 2008 SIAM International Conference on Data Mining*. 2008; 455–465.

Hotelling H. The generalization of Student's ratio. *Ann Math Statist*. 1931;2:360–378.

Hu H, Li J, Wang H, Daggard G, Shi M. A maximally diversified multiple decision tree algorithm for microarray data classification. In: *Workshop on Intelligent Systems for Bioinformatics*, 4 Dec 2006, Hobart, Australia.

Hubert L, Arabie P. Comparing partitions. *J Classification*. 1985;2:193–218.

Jeffery IB, Higgins DG, Culhane AC. Comparison and evaluation of methods for generating differentially expressed gene lists from microarray data. *BMC Bioinformatics*. 2006;7: 359.

Jiang W, Simon R. A comparison of bootstrap methods and an adjusted bootstrap approach for estimating the prediction error in microarray classification. *Statistics Med*. 2007;26:5320–5334.

Jones HL. Investigating the properties of a sample mean by employing random subsample means. *JASA*. 1956;51:54–83.

Juntu J, Jan Sijbers J, De Backer S, Rajan J, Van Dyck D. Machine learning study of several classifiers trained with texture analysis features to differentiate benign from malignant soft-tissue tumors in T1-MRI images. *J Magn Reson Imaging*. 2010;31:680–689.

Karniski W, Blair RC, Snider AD. An exact statistical method for comparing topographic maps. *Brain Topogr*. 1994;6:203–210.

Kerr MK, Churchill GA. Bootstrapping cluster analysis: assessing the reliability of conclusions from microarray experiments. *Proc Natl Acad Sci USA.* 2001;98:8961–8965.

Kishino H, Waddell PJ. Correspondence analysis of genes and tissue types and finding genetic links from microarray data. *Genome Inform.* 2000;11:83–95.

Korn E, Troendle JF, Mcshane LM, Simon R. Controlling the number of false discoveries: application to high-dimensional genomic data. *J Statist Planning Inference.* 2004;124:379–398.

Kuo W-J, Chang R-F, Moon WK, Lee CC, Chen D-R. Computer-aided diagnosis of breast tumors with different US systems. *Acad Radiol.* 2002;9:793–799.

Ladanyi A, Sher AC, Herlitz A, Bergsrud DE, Kraeft SK, Kepros J, McDaid G, Ferguson D, Landry ML, Chen LB. Automated detection of immunofluorescently labeled cytomegalovirus-infected cells in isolated peripheral blood leukocytes using decision tree analysis. *Cytometry Part A.* 2004;58A:147–156.

Lahiri SN. *Resampling Methods for Dependent Data.* New York: Springer; 2003.

Lee SMS, Pun MC. On *m* out of *n* bootstrapping for nonstandard *m*-estimation with nuisance parameters. *JASA.* 2006;101:1185–1197.

Lele S, Richtsmeier JT. Euclidean distance matrix analysis: confidence intervals for form and growth differences. *Am J Phys Anthropol* 1995;98:73–86.

Li C, Wong WH. Model-based analysis of oligonucleotide arrays: expression index computation and outlier detection, *Proc Natl Acad Sci USA.* 2001;98:31–36.

Lin M, Wei LJ, Sellers WR, Lieberfarb M, Wong WH, Li C. dChipSNP: significance curve and clustering of snp-array-based loss-of-heterozygosity data. *Bioinformatics.* 2004;20:1233–1240.

Lin S, Rogers JA, Hsu JC. A confidence-set approach for finding tightly linked genomic regions. *Am J Hum. Genet.* 2001;68:1219–1228.

Long AD, Mangalam HJ, Chan BY, Tolleri L, Hatfield GW, Baldi P. Improved statistical inference from DNA microarray data using analysis of variance and a Bayesian statistical framework. Analysis of global gene expression in *Escherichia coli* K12. *J Biol Chem.* 2001;276:19937–19944.

Ma S. Empirical study of supervised gene screening. *BMC Bioinformatics.* 2006;7: 537.

MacQueen JB. Some methods for classification and analysis of multivariate observations, *Proceedings of 5th Berkeley Symposium on Mathematical Statistics and Probability*, Berkeley, University of California Press, 1967;1:281–297.

Makinodan T, Albright JW, Peter CP, Good PI, Hedrick ML. Reduced humoral activity in long-lived mice. *Immunology.* 1976;31:400–408.

Mantel N. The detection of disease clustering and a generalized regression approach. *Cancer Res.* 1967;27:209–220.

MAQC Consortium: The MicroArray Quality Control (MAQC) project shows inter- and intraplatform reproducibility of gene expression measurements. *Nat Biotechnol.* 2006;24:1151–1161.

McIntyre LM, Martin ER, Simonsen KL, Kaplan NL. Circumventing multiple testing: a multilocus Monte Carlo approach to testing for association. *Genet Epidemiol.* 2000;19:18–29.

McLachlan G, Krishnan T. *The EM Algorithm and Extensions.* Hoboken, NJ: Wiley; 1996.

Milligan GW, Cooper MC. An examination of procedures for determining the number of clusters in a data set. *Psychometrika.* 1985;50:159–179.

Mootha VK, Lindgren CM, Eriksson K-F, Subramanian A, Sihag S, Lehar J, Puigserver P, Carlsson E, RidderstrAAle M, Laurila E, Houstis N, Daly MJ, Patterson N, Mesirov JP, Golub TR, Tamayo T, Spiegelman B, Lander ES, Hirschhorn JN, Altshuler D, Groop LC. PGC-1#-responsive genes involved in oxidative phosphorylation are coordinately downregulated in human diabetes. *Nat Genet.* 2003;34:267–273.

Mukherjee S. *Classifying Microarray Data Using Support Vector Machines, Understanding and Using Microarray Analysis Techniques: A Practical Guide.* Boston: Kluwer Academic Publishers; 2003.

Nichols TE, Holmes AP. Nonparametric permutation tests for functional neuroimaging: a primer with examples. *Hum Brain Mapping.* 2001;15:1–25.

Nichols TE, Holmes AP. Nonparametric permutation tests for functional neuroimaging. In: *Human Brain Function,* 2nd ed. R Frackowiak, KJ Friston, C Frith, R Dolan, Eds. New York: Academic Press; 2003.

Ogden RT, Tarpey T. Estimation in regression models with externally estimated parameters. *Biostatistics.* 2006;7:115–129.

Pang H, Kim I, Zhao H. Pathway-based methods for analyzing microarray data. In: *Analysis of Microarray Data.* Emmert-Sterib F, Dehmer M, Eds. Hoboken, NJ: Wiley-VCH; 2008.

Pantazis D, Nichols TS, Baillet S, Leahy RM. Spatiotemporal localization of significant activation in MEG using permutation tests. *Proceedings of the 18th Conference on Information Processing in Medical Imaging.* Springer Lecture Notes in Computer Science. 2003.

Pantazis D, Nichols TS, Baillet S, Leahy RM. A comparison of random field theory and permutation methods for the statistical analysis of MEG data. *NeuroImage.* 2005;25:383–394.

Parodi S, Izzotti A, Muselli M. Re: The central role of receiver operating characteristic (ROC) curves in evaluating tests for the early detection of cancer. *J Natl Cancer Inst*. 2005;97:234–235.

Parodi S, Pistoia V, Muselli M. Not proper ROC curves as new tool for the analysis of differentially expressed genes in microarray experiments. *BMC Bioinformatics*. 2008;9: 410.

Parodi S, Muselli M, Fontana V, Bonassi S. ROC curves are a suitable and flexible tool for the analysis of gene expression profiles. *Cytogenet Genome Res*. 2003;101:90–91.

Pei Y. Diagnostic approach in autosomal dominant polycystic kidney disease. *Clin J Am Soc Nephrol*. 2006;1:1108–1114.

Pepe MS, Longton G, Anderson GL, Schummer M. Selecting differentially expressed genes from microarray experiments. *Biometrics*. 2003;59:133–142.

Pesarin F. *Multivariate Permutation Tests*. Hoboken, NJ: Wiley; 2001.

Potter D, Griffiths D. Omnibus permutation tests of the overall null hypothesis in datasets with many covariates *J Biopharm Statist*. 2006;16:327–341.

Quinlan JR. Bagging, boosting, and C4.5. *Proceedings of the* 13th *National Conference on Artificial Intelligence*. American Association of Artificial Intelligence, Portland, OR. 1996; pp. 725–730.

Quinlan JR, Cameron-Jones RM. Induction of logic programs: FOIL and related systems. *New Generation Computing, Special issue on Inductive Logic Programming*. 1995;13:287–312.

Qiu X, Xiao Y, Gordon A, Yakovlev A. Assessing stability of gene selection in microarray data analysis. *BMC Bioinformatics*. 2006;7: 50.

Quackenbush J. Computational analysis of microarray data. *Nat Genet*. 2001;2:418–427.

Reiczigel, J. Bootstrap tests in correspondence analysis. *Appl Stochastic Models Data Anal*. 1996;12:107–117.

Reiner A, Yekutieli D, Benjamini Y. Identifying differentially expressed genes using false discovery rate controlling procedures. *Bioinformatics*. 2003;19:368–375.

Romano JP, Shaikh AM, Wolf M. Control of the false discovery rate under dependence using the bootstrap and subsampling (with discussion). *Test*. 2008;17:417–442.

Schall R. Assessment of individual and population bioequivalence using the probability that bioavailabilities are similar. *Biometrics*. 1995;51:615–626.

Schetinin V, Schult J. The combined technique for detection of artifacts in clinical electroencephalograms of sleeping newborns. *IEEE Trans Inf Technol Biomed*. 2004;8:28–35.

Sekihara K, Sahani M, Nagarajan SS. A simple nonparametric statistical thresh-olding for MEG spatial-filter source reconstruction images. *NeuroImage.* 2005;27:368–376.

Shapleske J, Rossell SL, Chitnis XA, Suckling J, Simmons A, Bullmore ET, Woodruff PTR, David AS. A computational morphometric MRI study of schizophrenia: effects of hallucinations. *Cerebral Cortex.* 2002;12:1331–1341.

Shoemaker JS, Lin SM. *Methods of Microarray Data Analysis*, IV. New York: Springer; 2005.

Sohn I, Owzar K, George SL, Kim S, Jung SH. A permutation-based multiple testing method for time-course microarray experiments. *BMC Bioinformatics.* 2009;10: 336.

Stewart C, Tsai CL, Roysam B. The dual-bootstrap iterative closest point algorithm with application to retinal image registration. *IEEE Trans Med Imaging.* 2003;22:1379–1394.

Stiglic G, Bajgot M, Kokol P. Gene set enrichment meta-learning analysis: next-generation sequencing versus microarrays. *BMC Bioinformatics.* 2010;11: 176.

Storey JD, Xiao W, Lee JT, Tompkins RG, Davis RW. Significance anal-ysis of time course microarray experiments. *Proc Natl Acad Sci USA.* 2005;102:12837–12842.

Subramanian A, Tamayo P, Mootha VK, Mukherjeed S, Eberta BL, Gillette MA, Paulovich A, Pomeroy SL, Golub TR, Lander ES, Mesirova JP. Gene set enrichment analysis: a knowledge based approach for interpreting genome-wide expression profiles. *Proc Natl Acad Sci USA.* 2005;102:15545–15550.

Suckling J, Davis MH, Ooi C, Wink AM, Fadili J, Salvador R, Welchew D, Sendur L, Maxim V, Bullmore ET. Permutation testing of orthogonal factorial effects in a language-processing experiment using fMRI. *Hum Brain Mapping.* 2006;27:425–433.

Suzuki R, Shimodaira H. An application of multiscale bootstrap resampling to hierarchical clustering of microarray data: How accurate are these clusters? *The Fifteenth International Conference on Genome Informatics.* 2004; P034.

Tan Q, Brusgaarda K, Torben A, Krusea TA, Edward Oakeley E, Hemmings B, Beck-Nielsena H, Hansenc L, Gastera M. Correspondence analysis of microarray time-course data in case—control design. *J Biomed Informatics.* 2004;37:358–365.

Thompson PM, Cannon TD, Narr KL, Erp T, Poutanen V-P, Huttunen M, Lönnqvist J, Standertskjfld-Nordenstam C-G, Kaprio J, Khaledy M, Dail R, Zoumalan CI, Toga AW. Genetic influences on brain structure. *Nat Neurosci.* 2001;4:1253–1258.

Tian L, Greenberg SA, Kong SW, Altschuler J, Kohane IS, Park PJ. Discovering statistically significant pathways in expression profiling studies. *Proc Natl Acad Sci USA*. 2005;102:13544–13549.

Troendle JF. A stepwise resampling method of multiple hypothesis testing. *JASA*. 1995;90:370–378.

Troyanskaya O, Garber ME, Brown PO, Botstein D, Altman RB. Nonparametric methods for identifying differentially expressed genes in microarray data. *Bioinformatics*. 2002;18:1454–1461.

Trujillano J, Sarria-Santamera A, Esquerda A, Badia M, Palma M., March J. Aproximacion a la metodologia basada en arboles de decision (CART). Mortalidad hospitalaria del infarto agudo de miocardio. *Gac Sanit*. 2008;22:65–72.

Tusher VG, Tibshirani R, Chu G. Significance analysis of microarrays applied to the ionizing radiation response. *Proc Natl Acad Sci USA* 2001;98:5116–5121.

van der Laan MJ, Bryan J. Gene expression analysis with the parametric bootstrap. *Biostatistics*. 2001;2:445–461.

Volkow ND, Rosen B, Farde L. Imaging the living human brain: magnetic resonance imaging and positron emission tomography. *Proc Natl Acad Sci USA*. 1997;94:2787–2788.

Wendel K, Väisänen O, Malmivuo J, Gencer NG, Vanrumste B, Durka P, Magjarevi R, Supek S, Pascu ML, Fontenelle H, Grave de Peralta Menendez R. EEG/MEG source imaging: methods, challenges, and open issues. *Comput Intell Neurosci*. 2009; Article ID 656092.

Wheldon MC, Anderson MJ, Johnson BW. Identifying treatment effects in multi-channel measurements in electroencephalographic studies: multivariate permutation tests and multiple comparisons. *Aust New Zealand J Statistics*. 2007;49:397–413.

Wigginton JE, Cutler DJ, Abecasis. A note on exact tests of Hardy–Weinberg equilibrium. *Am J Hum Genet*. 2005;76:887–883.

Yang C, Medioni G. Object modelling by registration of multiple range images. *Image Vision Comp*. 1992;10:145–155.

AUTHOR INDEX

Abecasis, G.R., 26, 173
Ai-Jun, Y., 70, 165
Albright, J.W., 169
Alter, O., 130, 165
Altman, R.B., 72, 173
Altschuler, J., 172
Altshuler, D., 170
Amtmann, A., 166
Anagnostopoulos, D., 167
Anderson, G.L., 171
Anderson, M.J., 35, 173
Andreasen, N.C., 165
Arabie, P., 94, 168
Armengaud, P., 166
Arndt, S., 35, 165
Aslandogan, Y.A., 72, 166
Augustin, T., 165

Badia, M., 173
Bailey, J., 105, 126, 168
Baillet, S., 170
Bair, E., 129–130, 165
Bajgot, M., 133–135, 172
Baldi, P., 169
Beck-Nielsena, H., 172

Bellec, P., 85, 165
Belmonte, M., 36, 165
Benali, H., 85, 165
Benjamini, Y., 61, 69, 167, 171
Bergsrud, D.E., 169
Bhavsar, P., 166
Biscay, R., 167
Bittner, M., 51, 166
Blair, R.C., 30, 35, 168
Bolles, R.C., 96, 167
Bonassi, S., 171
Bosch, J., 168
Botstein, D., 72, 130, 135, 165, 167, 168, 173
Boulesteix, A.-L., 165
Bourgon, R., 58, 166
Breitling, R., 53, 166
Brors, B., 24, 166
Brown, P.O., 72, 130, 135, 165, 167, 173
Brozzi, A., 166
Brusgaarda, K., 172
Bryan, J., 99, 173
Bühlmann, P., 24, 167
Bullmore, E.T., 36, 171, 172

Analyzing the Large Numbers of Variables in Biomedical and Satellite Imagery, First Edition.
Phillip I. Good.
© 2011 John Wiley & Sons, Inc. Published 2011 by John Wiley & Sons, Inc.

SUBJECT INDEX

Analyzing the Large Numbers of Variables in Biomedical and Satellite Imagery, First Edition.
Phillip I. Good.
© 2011 John Wiley & Sons, Inc. Published 2011 by John Wiley & Sons, Inc.